Springer Theses

Recognizing Outstanding Ph.D. Research

Aims and Scope

The series "Springer Theses" brings together a selection of the very best Ph.D. theses from around the world and across the physical sciences. Nominated and endorsed by two recognized specialists, each published volume has been selected for its scientific excellence and the high impact of its contents for the pertinent field of research. For greater accessibility to non-specialists, the published versions include an extended introduction, as well as a foreword by the student's supervisor explaining the special relevance of the work for the field. As a whole, the series will provide a valuable resource both for newcomers to the research fields described, and for other scientists seeking detailed background information on special questions. Finally, it provides an accredited documentation of the valuable contributions made by today's younger generation of scientists.

Theses are accepted into the series by invited nomination only and must fulfill all of the following criteria

- They must be written in good English.
- The topic should fall within the confines of Chemistry, Physics, Earth Sciences, Engineering and related interdisciplinary fields such as Materials, Nanoscience, Chemical Engineering, Complex Systems and Biophysics.
- The work reported in the thesis must represent a significant scientific advance.
- If the thesis includes previously published material, permission to reproduce this must be gained from the respective copyright holder.
- They must have been examined and passed during the 12 months prior to nomination.
- Each thesis should include a foreword by the supervisor outlining the significance of its content.
- The theses should have a clearly defined structure including an introduction accessible to scientists not expert in that particular field.

More information about this series at http://www.springer.com/series/8790

Patrick L.S. Connor

Inclusive *b* Jet Production in Proton-Proton Collisions

Precision Measurement with the CMS
experiment at the LHC at $\sqrt{s} = 13$ TeV

Doctoral Thesis accepted by
The University of Hamburg, Hamburg, Germany

 Springer

Author
Dr. Patrick L.S. Connor
CMS
Deutsche Elektronen-Synchrotron DESY
Hamburg, Germany

Supervisor
Dr. Priv.-Doz. Hannes Jung
CMS
The Deutsches Elektronen-Synchrotron
Hamburg, Germany

ISSN 2190-5053   ISSN 2190-5061   (electronic)
Springer Theses
ISBN 978-3-030-34385-9   ISBN 978-3-030-34383-5   (eBook)
https://doi.org/10.1007/978-3-030-34383-5

This Springer imprint is published by the registered company Springer Nature Switzerland AG
The registered company address is: Gewerbestrasse 11, 6330 Cham, Switzerland

« *La pensée*
ne doit jamais
se soumettre,
ni à un dogme,
ni à un parti,
ni à une passion,
ni à un intérêt,
ni à une idée préconçue,
ni à quoi que ce soit,
si ce n'est
*aux **faits** eux-mêmes,*
parce que,
pour elle,
se soumettre,
ce serait cesser d'être. »
Henri POINCARÉ

Speech for the 75th anniversary
of the *Université libre de Bruxelles*
21 November 1909

给王群

Supervisor's Foreword

Measurements at the Large Hadron Collider (LHC) at CERN are a very rich source for particle physics, and many new phenomena could be studied for the very first time. A highlight of the measurements at the LHC was discovery of the Higgs boson in 2012. Apart from the discovery, the understanding of interactions between high-energetic particles has reached a new and fascinating level: jets, collimated streams of particles, have been measured up to highest transverse momenta of a few TeV, a region which was never accessible before. At these high energies, the jets become more and more pencil-like. Since these high transverse momentum jets can be measured very precisely, measurements of correlations become possible, which one could not dream of before.

The understanding of the production mechanism of the Higgs boson and any other process requires deep knowledge of the basic interactions of proton-proton collisions: the cross sections are calculated as a convolution of the probability to find a parton in one of the protons with the probability to find another parton in the other proton and the probability that both partons interact with each other. While the basic idea is known since long, much progress has been achieved in the calculation including higher orders in perturbation theory. An extension of this basic idea was proposed, which includes a further degree of freedom into the calculations: the transverse momentum of initial partons.

With the measurements at the LHC our knowledge of fundamental interactions has increased enormously in the region of highest energies but also in the region of smallest momenta of the interacting partons. In this respect, LHC is a machine of the extremes.

This thesis provides the first measurement of heavy-flavour tagged jets up to highest energies. Never before was it possible to measure bottom tagged jets in the TeV region. At these high energies, the tagging of bottom jets becomes extremely difficult and a special effort had to be taken to perform a reliable measurement, going much beyond what was normally available. The results of the measurement were surprising: the production cross section of bottom tagged jets at high transverse momentum came out to be significantly larger than expected from calculations of direct bottom production. It turned out that at these high transverse momenta, often bottom quarks are produced in the showering process, initiated from light partons. Calculations which include bottom production in the parton cascade are in very good agreement with the measurements.

With this first measurement of bottom jets in the high transverse momentum region, a new field was opened: a systematic comparison of flavour tagged jets with inclusive jets over the whole kinematic region became possible with the possibility to study in detail the flavour blindness of strong interactions, when the mass thresholds become unimportant. The rather constant ratio of the measured bottom tagged to inclusive jets supports this hypothesis impressively.

Particle physics is one of those enterprises that can only be performed by collaborations that bring together different people with different skills and ideas, from very different regions of our world. It is a real privilege to be part of this scientific, but also socially and culturally enriching environment. Particle physics is one example of how global and peaceful collaboration can achieve aims that seem unthinkable otherwise. In the environment of a diverse working group and institute, Patrick performed his research and obtained his Ph.D. degree.

It is a real pleasure to write this foreword for an excellent thesis and it makes me happy to see that Patrick found also new friends during his Ph.D. and attracts new people to the field.

Hamburg, Germany   Hannes Jung
April 2019

Abstract

A measurement of the double differential cross section for inclusive b jet production in proton-proton collisions as well as fraction of b jets in the inclusive jet production is presented as a function of the transverse momentum p_T and the absolute rapidity $|y|$. The data samples were collected in the CMS experiment at LHC during 2016 and correspond to an integrated luminosity of $35.2\,\mathrm{fb}^{-1}$ at a centre-of-mass energy of 13 TeV. The jets are selected with $p_T > 74\,\mathrm{GeV}$ and $|y| < 2.4$; the b jets must contain a B hadron. The measurement has significant statistics up to $p_T \sim \mathcal{O}(\mathrm{TeV})$. Advanced methods of unfolding are performed to extract the signal. It is found that fixed-order calculations with underlying event describe the measurement well.

Acknowledgements

First of all, my gratitude goes to Hannes JUNG, who supervised my work during three and a half years. Hannes, this time, was really grand: always reachable, always supporting, always positive. Besides physics, you also taught me how to behave in a group (science is also made of social interactions). This work together will for sure have a great impact in my whole life.

Many thanks to Radek, Paolo and Benoît who, as post-docs in the group, largely contributed to my work; now it is my turn to take over. Ola, Armando, Daniela, Juan, Jindrich, Samantha, Svenja, Engin: thanks for all the good time spent together! (And by the way: thanks for the hat!).

But DESY is not limited to the QCD group. First, I spent also a lot of time in the alignment group: Matthias, Rainer, Nazar, Valeria, Gregor, Rostyslav, Chayanit, Claus, all of them have also contributed actively to my education. The DESY staff members, also, have always kindly helped me even on short notice: Achim, Stefan Schmidt, Olaf, Katerina, Christophe, Matthias KASEMANN; for their time and help, I want to thank them too.

Merci à James, mon scientifique irlandais favori, qui a relu mon anglais dans quelques chapitres de cette thèse.

Grazie a Matteo, il mio personale esperto del b tagging, che a avuto la buon'idea di lavorare nello stesso ufficio accanto a me. Thanks also to Kirill SKOVPEN, Ivan MARCHESINI and Caroline COLLARD, for their precious help in b tagging.

Au CERN, merci à Olivier BONDU et Gaëlle BOUDOUL, en particulier, qui m'ont bien aidé lors de mes premiers pas dans le monde du Tracker DQM.

In Helsinki, thanks to Mikko VOUTILAINEN and Tapio LAMPÉN, who helped me in their respective fields.

Hartelijk dank, Hans: elke keer is het weer tof, samen te diskuteren.

Even though from ATLAS, I would like to thank Nataliia ZAKHARCHUK and Jihyun JEONG for their friendship, as well as So Young Shim and Shruti Patel from the theory group.

Ich bedanke mich auch bei Birgit & Gabriele, die immer so hilfbereit gewesen sind, und bei den Pförtnern, die immer sehr nett waren, als ich den Schlüßel des Flügels abholen kam.

Thanks to Samantha, who, first, helped me in the quest of a flat in Hamburg. Merci Marisa, pour l'ex-future colocation! Tu auras été mon premier contact à Hambourg en dehors de DESY. Danke, Martin, für die tolle WG in Eimsbüttel. Danke, Nils, Karin, Nora, Ruth, Marc-Ol für die WG in Altona. Je remercie la famille BOLLIGER, à commencer par Aline qui m'a obtenu une chambre chez ses parents à Genève pour trois mois. Und danke, Annette & David für die WG in Bahrenfeld.

Ohne die Musik wären die Arbeit und das Leben nicht so schön: Deswegen bedanke ich mich auch bei meinem Chor *hamburgVOKAL* und bei meinen Musiklehrern Rainer und Michèle. Danke, auch, an Xaver, der mir umsonst Musiktheorie unterrichtet hat.

Außer der Musik habe ich auch viele Energie beim ETV Judo verwendet: Danke Mana, Stefan, Klara, Jawid und Eki!

Je ne pourrais pas me permettre d'oublier mon promoteur de mémoire, Laurent FAVART, sans qui il n'y aurait eu ni Hambourg ni DESY pour moi. À l'IIHE, merci à David, Hugo, Isabelle, Hugues, Nastja, ... De l'*Université libre de Bruxelles*, deux professeurs en particulier ont contribué au dévelopement de mon intérêt pour la physique fondamentale: Thomas HAMBYE & Pierre GASPARD.

Remontant encore plus loin dans le temps, merci à Martin CASIER, qui a déclenché la première étincelle d'intérêt pour la physique. De tout coeur, merci à Antoine HOEFFELMAN ainsi qu'à tous les membres de l'*Astroguindaine*. Hervé, quel plaisir de t'avoir vu à Hambourg pour ma défense.

Je pense aussi à mes professeurs du *Lycée Émile Jacqmain*, en particulier à Paul VAN LANGENHOVEN & Rony VINDELYNCKX qui, plus que tout autre, m'ont appris à structurer un texte. Leur trace dans ce volume est indéniable.

Bien entendu, mes derniers remerciements, les plus sincères également, vont à ma famille, surtout mes parents. Et merci à toi, ma Qun, du plus profond de mon coeur, pour avoir rejoint mes jours terrestres. 我爱你, 王群!

Preamble

More than any other field of science, the complexity of the theory and the size of the experiments in Particle Physics require a wide knowledge of many different topics as well as large collaborations to face the unprecedented amount of data to acquire and to analyse. The training of the physicist comprises maths, fundamental physics, detector physics, modelling, data reduction and programming. Today's collaborations gather up to a few thousands of analysts and engineers per experiment; 50 years were needed between the theoretical prediction of the so-called Higgs boson and its experimental discovery; around 200 papers were submitted when, in 2016, a new resonance was believed to appear around 600 in the invariant mass spectrum of the diphoton production. One sometimes evokes Particle Physics as the physics of the extremes; I think this does not only have to do with the quantum and relativistic scales.

From October 2014 to March 2018, I had the chance to participate to various aspects of research in Particle Physics under the supervision of Hannes JUNG. Initially, following up on the main topic of my master thesis, I was supposed to work on the DY process and associated jet production, and started some MC investigations. But the work really first started with contributions to the measurement of the MB cross section at LHC Run-II. In parallel, I started my service tasks in the group dedicated to the alignment of the tracker system at CMS, taking advantage of the strong involvement of the DESY group in this area; I continued this activity throughout the whole duration of the doctorate. After the minimum bias analysis, we decided together with Hannes Jung to change the main topic of the thesis to b jet measurements, more appropriate with respect to the other activities in the group since the departure of a member of our group with whom I was supposed to work. Initially, the new project consisted in the investigation of the associated jet production to $b\bar{b}$ production, starting with "some quick analysis on the inclusive b jet production." It turned out, however, that the b jet production at the TeV scale was non-trivial, and became the main topic of the thesis; this will be the object of the present memoir. Finally, in parallel of this work, I redeveloped from scratch the `TMDplotter`, an on-line facility for plotting parton densities; I also participated as a teaching assistant the exercise sessions of the MC lectures by Hannes Jung, in Hamburg and in Antwerp; and I took an active part in the organisation of internal paper reading sessions at DESY.

The present thesis is organised in three parts. The first part covers a general introduction: historical, theoretical and experimental; running across this different aspects, it aims at offering a general understanding of the stakes in Particle Physics, together with a motivation of the study of these famous b jets. Here, the aim is not to substitute to manuals, but to draw up the main lines of Particle Physics in a more complete way than a list of references and to introduce some concepts and formulae —maybe a bit more—that will be useful in the following. Benefiting from the concepts introduced in the first part, and following a logic of going from the top to the bottom, the second part will treat the analysis of data at CMS from the LHC taken in 2016; it will be more technical and specific, representing the bulk of the work that I have performed along this doctorate in the topic of b jets. Questions regarding various calibrations, modern investigations in b tagging and advanced unfolding techniques will be addressed there; the literature about these not always being easy to approach, the detail in which they are described in this thesis was also reached in the hope and the wish of being useful to future students. In addition to these two parts, a third part, much shorter, will be dedicated to discussions about our current knowledge about b jets, and prospects about future measurements and predictions; finally, an appendix will close off this thesis, describing activity in tracker alignment, where I have been strongly involved all along my *Promotion* at DESY in parallel of the topic presented in this thesis.

Although completely transparent through this thesis, another aspect of great impact in modern Particle Physics is the ever increasing need in computing skills, essential to treat large amount of data in an efficient way and to collaborate with other physicists. It is not a coincidence if the modern World Wide Web was developed at CERN in the context of Particle Physics in 1991. Advanced programming skills are nowadays essential to conduct an analysis successfully, and since schools and universities still rarely offer their proper teaching, it important to stress that a significant part of the time spent before and at DESY had to be dedicated to their self-learning, especially to that of C++.

Lastly, before getting down to the business, I would make a mistake if I would not mention here a non-scientific but nonetheless non-negligible aspect of working in a large collaboration such as the ones in Particle Physics. Indeed, since it gathers physicists from all around the world, Particle Physics is also made of social interactions, together with its cultural shocks; the social aspect of the work in collaboration has three entangled consequences. The first consequence is that politics is a significant component of research in Particle Physics, sometimes more driving choices than scientific arguments; as much as I have been able to do so, I have tried to avoid politics and attempted to follow the philosophy of my *Alma Mater*, the *Université libre de Bruxelles*, very well summarised in Poincaré's quotation given in the first pages.[1] The second one is that this thesis was to be written in the self-proclaimed international language, English, although by a Belgian graduating in Germany, having therefore *a priori* few to do with this language (leaving aside the Irish name...); on this point, I would like to mention that I chose to use the European spelling rather than the American, which, as a corollary of the previous point, is a political choice. The third and last point is that the confrontation of cultures implies the confrontation of ideas and approaches, not only in science but also in daily life; in my case, it has first implied living in Germany and learning its language. This has for sure influenced me, and if I had been to write this thesis in Belgium, it would likely look quite different. But more than anything else—and I will finish the preamble there—it has also meant meeting my fiancée, Wang Qun; and despite the great interest that I have had for the topic of the present volume, our encounter is certainly the most important thing that Particle Physics will have ever granted to me.

---

[1]Translation to English: Thinking must never submit itself neither to a dogma, nor to a political party, nor to a passion, nor to an interest, nor to a preconceived idea, nor to anything except facts themselves, because for it to submit would mean cease existing.

Contents

Part III Conclusions

Acronyms

Geant4	GEometry ANd Tracking
NLOJet++	NLO Jet in C++
Pluto	After the magnet of the detector[2]
Professor	PROcedure For EStimating Systematic errORs
Zeus	in reference to the relation of Zeus and Hera in the mythology
CSVv2	Combined-Secondary-Vertex
CUETHppS1	CMS UE Tune HERWIG++ Set 1
CUETP8M1	CMS UE Tune PYTHIA 8 Monash 1
JP	Jet Probability
NaN	Not a Number
PXB	PiXel Barrel
PXF	PiXel Forward
TEC	Tracker End-Caps
TIB	Tracker Inner Barrel
TID	Tracker Inner Disk
TOB	Tracker Outer Barrel
cMVAv2	combined-Multi-Variate-Analysis
ABM	After the three physicists ALEKHIN, BLÜMLEIN and MOCH
ALICE	A Large Ion Collider Experiment
ARGUS	A Russian-German-United States-Swedish collaboration[3]
ATLAS	A Toroidal LHC ApparatuS
AVF	Adaptive Vertex Fitter
AVR	Adaptive Vertex Reconstruction
BaBar	$B\bar{B}$
BBR	Beam-Beam Remnants
BDT	Boosted-Decision Tree
BEH	After the three physicists BROUT, ENGLERT and HIGGS
Belle	French word for beauty

---

[2]Aachen-DESY-Hamburg-Siegen-Wuppertal-München collaboration.

[3]Though joined later by other countries.

BFKL	After the four physicists BALITSKY, FADIN, KURAEV and LIPATOV
BLT	Bottom Line Test
Booster	Proton Synchrotron Booster
BR	Branching Ratio
BSM	Searches Beyond the SM
CASTOR	Centauro And STrange Object Research in nucleus-nucleus collisions at the LHC
CCFM	After the four physicists CATANI, CIAFALONI, FIORANI and MARCHESINI
CDF	Collider Detector at Fermilab
CERN	European Organisation for Nuclear Research[4]
CESR	Cornell Electron-positron Storage Ring, pronounced "Caesar"
CKM	After the three physicists CABIBBO, KOBAYASHI and MASKAWA
CLEO	*Cleopatra* (for her/its proximity with Caesar)
c.m.s.	Centre-of-mass
CMS	Compact Muon Solenoid
CSC	Cathode Strip Chamber
CT	Closure Test
CTEQ	Coordinated Theoretical-Experimental Project on QCD
CUSB	Columbia University-Stony Brook
DASP2	Double Arm SPectrometer[5]
DESY	*Deutsches Electroknen-Synchrotron*
DGLAP	After the five physicists DOKSHITZER, GRIBOV, LIPATOV, ALTARELLI and PARISI
DHHM	DESY-Hamburg-Heidelberg-München collaboration
DIS	Deeply Inelastic Scattering
DØ	After the location of the detector
DORIS	*DOppel-RIng Speicher*
DPS	Double-Parton Scattering
DT	Drift Tube
DY	Drell-Yan
EB	ECAL Barrel
ECAL	Electromagnetic CALorimeter
EE	ECAL End-caps
e.m.	electromagnetic
EOM	Equations of Motion
EW	Electroweak
FCR	Flavour Creation
FEX	Flavour Excitation
FF	Fragmentation Function
FGR	Fermi Golden Rule

---

[4]Originally *Conseil Européen pour la Recherche Nucléaire*, kept for the proximity with the Germanic root *kern*, meaning "nucleus".

[5]DESY-Dortmund-Heidelberg-Lund.

FO	Fixed order
FSR	Final-State Radiation
GPMC	General-Purpose MC Event Generator
GSP	Gluon Splitting
H1	HERA-1
HB	HCAL Barrel
HCAL	Hadronic CALorimeter
HE	HCAL End-caps
HEP	High-Energy Physics
HERA	*Hadron-Elektron-RingAnlage*
HERA-B	HERA Beauty
HERAPDF	HERA PDF sets
HF	Heavy Flavour
HF	HCAL Forward
HLS	High-Level Structure
HLT	High-Level Trigger
HO	HCAL Outer
IC-PR	Iterative Cone algorithm with Progressive Removal procedure
IC-SM	Iterative Cone algorithm with Split Merge procedure
IP	Interaction point
IRC	Infrared and collinear safe
ISOLDE	Isotope Separator On Line DEvice
ISR	Initial-State Radiation
ISR	Intersecting Storage Rings
IVF	Inclusive Vertex Fitter
JEC	Jet Energy Correction
JER	Jet Energy Resolution
JES	Jet Energy Scale
KEK	Koo Energy Ken
KEKB	KEK Beauty
L1	Level 1
LEIR	The Low Energy Ion Ring
LEP	Large Electron Positron Collider
LHC	Large Hadron Collider
LHCb	LHC beauty
LHCf	LHC forward
LINAC 2	Linear Accelerator 2
LINAC 3	Linear Accelerator 3
LLN	Law of Large Numbers
LO	Leading Order
LS	Lumi Section
Mark	(unknown origin)
MB	Minimum Bias
MC	Monte Carlo
ME	Matrix Element

MET	Missing Transverse Energy
MMHT	After the names of the four physicists HARLAND-LANG, MARTIN and MOTYLINSKI, THORNE
MoEDAL	Monopole and Exotics Detector At the LHC
MPI	Multi-Parton Interaction
MSTW	After the names of the four physicists MARTIN, STIRLING, THORNE and WATT
MVA	Multi-Variate Analysis
n.d.f.	Number of degrees of freedom
NLO	Next to Leading Order
NNLO	Next to Next to Leading Order
NNPDF	Neural-Network PDF
p.d.f.	probability density function[6]
PDF	Parton Distribution Function
PDG	Particle Data Group
PEP	Positron-Electron Project
PF	Particle-Flow
PM	Probability Matrix
POCA	Point Of Closet Approach
pQCD	perturbative QCD
PS	Parton Shower
PS	Proton Synchrotron
PU	Pile-up
PV	Primary Vertex
QCD	Quantum Chromodynamics
QED	Quantum Electrodynamics
QFT	Quantum Field Theory
RM	Response Matrix
RPC	Resistive Plate Chamber
SF	Scale Factor
SIScone	Seedless and Infrared Safe Cone
SLAC	Stanford National Accelerator Laboratory
SLC	SLAC Linear Collider
SLD	SLAC Large Detector
SM	Standard Model
SPS	Super Proton Synchrotron
$Sp\bar{p}S$	Super Proton—Antiproton Synchrotron
SV	Secondary Vertex
SVD	Singular Value Decomposition
TMD	Transverse-Momentum-Dependent PDF
TOTEM	TOTal cross section, Elastic scattering and diffraction dissociation Measurement at the LHC
TRK	Tracker

---

[6]Not to be confused with PDF.

UA1	Underground Area 1, named after the location of the detector
UE	Underlying Event
WM	Weak Mode
WP	Working Point

Part I
Preliminaries

Chapter 1
Introduction to High Energy Physics

THE OBJECT OF HIGH ENERGY PHYSICS IS the study the constituents of matter and their interactions at quantic and relativistic scales, conditions only reachable with very high energies. At these scales, matter and interactions are both described in terms of *particles*.

In this chapter, we give a general introduction to HEP: the context, history, achievements and challenges of this field of research.

1.1 Introduction

1.1.1 Units

Being both relativistic and quantic, the scope of HEP is reflected in terms of units [1].

1. Electric charges are counted in units of the *elementary charge*:

$$e = 1.602,176,6208(98) \times 10^{-19} \, \text{C} \qquad (1.1)$$

2. Terms of the dynamical equations are of order of the *reduced Planck constant*[1]:

$$\hbar = \frac{h}{2\pi} = 1.054,571,800(13) \times 10^{-34} \, \text{Js} \qquad (1.2)$$

3. Velocities are measured in units of the speed of light[2]:

---

[1] $h = 6.626\,070\,040(81) \times 10^{-34}$ Js.

[2] The symbol c stands for *celerity*, an outmoded synonymous of velocity.

© Springer Nature Switzerland AG 2019
P. L. S. Connor, *Inclusive b Jet Production in Proton-Proton Collisions*, Springer Theses,
https://doi.org/10.1007/978-3-030-34383-5_1

$$c = 299{,}792{,}458 \,\text{m/s} \tag{1.3}$$

Units of energy are *electron-Volt*, or multiples:

$$1\,\text{GeV} = 1.602{,}176{,}6208(98) \times 10^{-10}\,\text{J} \tag{1.4}$$

1 eV corresponds to the kinetic energy that an electron gains when accelerated from rest in an electric potential of 1 V; moreover, 1 GeV roughly corresponds to the mass of the proton.[3] Momentum and mass are respectively measured in GeV/c and GeV/c^2; in practice however, the speed of light c is often omitted, as well as the $\hbar$ constant, both straightforward to recover into the formulae using dimensional analysis.

« *High Energy* » refers to the relation that can be derived using dimensional analysis between an energy E to a wavelength λ:

$$E = 2\pi \frac{\hbar c}{\lambda} \quad \text{or more simply} \quad E = \frac{2\pi}{\lambda} \tag{1.5}$$

i.e. probing high (low) energy scales implies probing low (high) distance scales.

1.1.2 Fundamental Interactions

Nowadays, nature is understood in terms of four *fundamental interactions*:

gravitation Gravitation describes interactions between objects due to their masses. Typical systems purely based on this force are solar systems, usually at large distance scales. Gravitation is by far the weakest interaction among the four; however, since masses are always positive, it is only cumulative and becomes therefore the dominant interactions from scales of $\mathcal{O}(1\,\text{m})$.

electromagnetism Electromagnetism describes interactions between objects carrying an electric charge (so far, no magnetic charge has been observed). It is indeed the dominant force at scales from $\mathcal{O}(10^{-10}\,\text{m})$ to $\mathcal{O}(10^{-3}\,\text{m})$. Typical structures holding via the electromagnetic interaction are atoms and molecules; it also explains γ decay, as well as gaseous, liquid and solid states of matter.

strong nuclear interaction The strong (nuclear) interaction describes interactions of components and subcomponents of the atomic nucleus. "Nuclear" refers to the scales at which they take place, below $\mathcal{O}(10^{-14})$; the only macroscopic manifestation is α decay (emission of a nucleus of helium). In general, the study of the strong interaction requires very specific set-ups, as will be the case in this thesis.

weak nuclear interaction Finally, the weak (nuclear) interaction takes place only at small distance scales, similarly to the strong interaction. It is responsible for β decay (emission of an electron), which is crucial in the nucleosynthesis of stars, but no analogous system like planetary systems, molecules or nuclei may

---

[3] $h = 938.272046(21)\,\text{MeV}/c^2$.

be found for holding only thanks to the weak interaction. This is related to the fact that similarly to the strong interaction, it takes only place at scales of the nucleus, but is much weaker than the strong interaction. However, unlike the strong interaction, it does not only affect the constituents of the nucleus but all particles of matter.

At the scale of experimentation of HEP, gravitation is too weak to produce any measurable effect; moreover, it is extremely difficult to formalise a quantum theory of gravitation. Nowadays, only the three other interactions are physically experimented in HEP and mathematically described within the so-called SM. The SM will be the object of the next chapter.

1.1.3 Particles

In general, any object that can be regarded as *pointlike* can be called a *particle*:

- galaxies in the universe (cosmology),
- stars in a galaxy or planets in a solar system (astrophysics),
- molecules in a medium (statistical physics)
- atoms in a molecule (chemistry),
- nucleus in an atom or nucleons in a nucleus (nuclear physics),
- and partons in a proton (particle physics).

But properly said, Particle Physics concerns the *fundamental* and *composite* particles, i.e. the tiniest components and the sets made of these. It is in this sense that *particles* shall here be meant.

Essentially, these particles have two peculiar behaviours:

- Most of them are unstable and decay in a very short time (at most a small fraction of a second—see Fig. 1.1). Normal matter only consists of atoms made of stables particles: the protons, neutrons and electrons.
- When two particles collide *violently*, they may *produce* other particles.

The study of HEP consists in trying to understand these two behaviours. In particular, in this thesis, we are going to study an unstable particle, the *b* quark, which can be produced by colliding protons.

1.2 History

In this section, we recall some key steps in the history of the discovery of the fundamental constituents of matter, following both chronology and decreasing distance scale, as shown in Table 1.1.

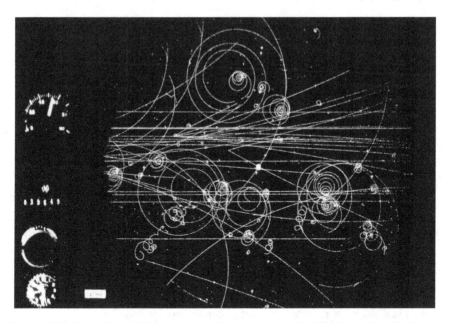

Fig. 1.1 This image from 1960 is of real particle tracks formed in CERN's first liquid hydrogen bubble chamber to be used in experiments. It was a tiny detector by today's standards at only 32 cm in diameter. Negatively charged pions with an energy of 16 GeV enter from the left. One of them interacts with a proton in the liquid hydrogen and creates sprays of new particles, including a neutral particle (a lambda) that decays to produce the "V" of two charged particle tracks at the centre. Lower-energy charged particles produced in the interactions spiral in the magnetic field of the chamber. The invention of bubble chambers in 1952 revolutionized the field of particle physics, allowing real particle tracks to be seen and photographed, after releasing the pressure that had kept a liquid above its normal boiling point. Figure reproduced with permission from European Organisation for Nuclear Research (CERN) [2]

Table 1.1 A few key figures relating the energy scale and the involved type of object. Note that 1 TeV represents approximately the kinetic energy carried by a domestic fly

Energy scale	Distance scale	Object
keV	10^{-10} m	Atom
MeV	10^{-13} m	Nucleus
GeV	10^{-16} m	Proton
TeV	10^{-19} m	Parton

1.2.1 Atomism (10^{-10} m)

The modern theory of atomism started with the publication of the book *Les Atomes* [3] in 1913 by Jean PERRIN, where thirteen different, compatible measurements of the *Avogadro number* were presented:

$$N_A \approx 6 \cdot 10^{23} \, \text{mol}^{-1} \tag{1.6}$$

This number is the typical number of atoms to be found in a few centimeters. Matter is not a *continuum* but is made of small corpuscles.

1.2.2 Nucleus (10^{-13} m)

The existence of the electron was admitted but no such oppositely charged particle had been found. The atom was thought to be a diffuse, positive body in which the electrons, negative, would shelter (*Thomson model* or *plum-pudding model*). Between 1908 and 1913, the *golden-foil* experiment [4] by Sir Ernest RUTHERFORD, Hans GEIGER and Ernest MARSDEN highlighted the existence of a charged point-like object—the *nucleus*—in atoms. The experiment (shown in Fig. 1.2) consists in bombing a golden foil gets with alpha rays, which can be found in naturally radioactive sources (like 238-uranium); according to the plum-pudding model, the radiation should have gone through the foil; however, they observed some alpha rays coming back. RUTHERFORD said: "It was almost as incredible as if you fired a 15-inch shell at a piece of tissue paper and it came back and hit you!" This was the first sign for the existence of a heavy, charged nucleus. The size of the nucleus was at most of the order of 10^{-13} m, since the energy of a natural source of alpha rays is around 5 MeV. This new model of the atom made of a small nucleus surrounded by an electronic cloud is called *Rutherford atom*.

1.2.3 Nucleus Structure (10^{-15} m)

From the 1950s at SLAC in the U.S. and later in the 1960s at DESY, the nucleus was probed with particle beams at an energy scale of the order of 100 MeV–1 GeV. Similarly to the golden-foil experiment, a nuclear target was bombed with a beam of electrons:

$$e + N \longrightarrow e + X \tag{1.7}$$

where e stands for *electron*, N for *nucleus* and X for some additional production. The nucleus itself was found to have a structure, make of pointlike *nucleons* (either protons or neutrons), arranged in a similar way to the structure of the electrons in the atom.

1.2.4 Nucleon Substructure (10^{-18} m)

At a still lower distance scale, i.e. with higher-energy beams, protons and neutrons also were found not to be pointlike, but with a substructure surprisingly different to

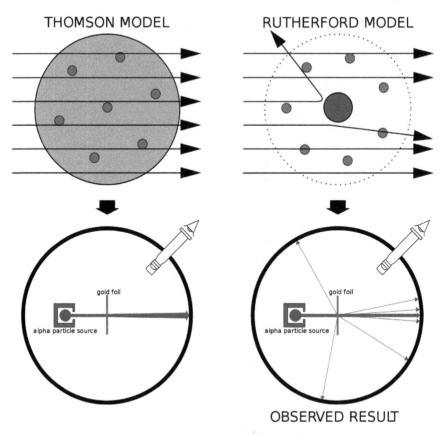

Fig. 1.2 A simple diagram illustrating the Geiger–Marsden experiment. The left column shows the scattering pattern that the experimenters expected to see, given the plum pudding model of the atom. The right column shows the actual results, along with Rutherford's new planetary model [5]

the atoms'. At that time, electrons were scattered on protons:

$$e + p \longrightarrow e + X \tag{1.8}$$

The results could be interpreted in two complementary ways:

1. the study the kinematics of the outgoing electron e led to the *parton model*, imagined by Richard FEYNMAN;
2. and independently the study of the symmetries of the hadronic production X led to the *quark* model, imagined by Murray GELL-MANN.

In the former, the *scattering* effects on the proton suggested pointlike subcomponents to exist, called *partons*, coherently moving without interacting with one another. In the latter, the existence of different subparticles, called *quarks*, could explain

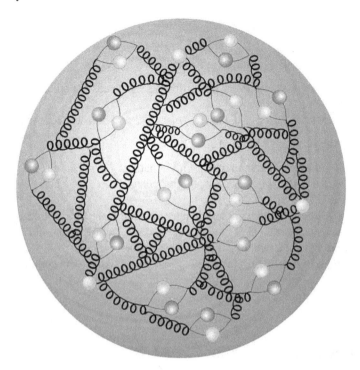

Fig. 1.3 On this artistic view, the proton is made of *quarks* (green), *antiquarks* (orange) and *gluons* (curly lines). There are three more quarks than antiquarks, called *valence quarks* (one can be found in the top, one in the right hand side, one in the bottom left). Quarks and antiquarks radiate gluons, themselves either radiating other gluons or decaying into a quark and an antiquark. *Source* DESY [6]

some *symmetries* among different types of particles. The two models were based on different observations and described the proton differently.

In the 1990s, at the HERA *ep* collider, at $E \sim 100\,\text{GeV} \to 10^{-18}$ m, the H1 and ZEUS collaborations measured the content of the proton in terms of partons [7] (Fig. 1.3).

Since then, despite active searches, no new substructure has been found. However, six different *flavours* of quarks have been found. This thesis is dedicated to the study of one of them: the *b* quark.

1.3 Experimentation

We now discuss the experimental possibilities in HEP: first how to achieve the right conditions of experimentation, and what can exactly be measured.

1.3.1 Sources

Experimentation in HEP is limited in two aspects:

- Since most particles have a very short lifetime, one needs to find or set up sources of particles.
- Since particles have a very small size, very specific detectors needs to be set up.

One may distinguish three types of sources:

radioactive elements   These can be found in nature or synthesised. This was how RUTHERFORD et al. first studied the atomic nucleus. Detection of particles are also performed.

cosmic rays   Stars radiate particles, which scatter on molecules in the atmosphere. The *cosmic microwave background* also gives a picture of the universe when atoms were not yet bound together.

scattering experiments   Accelerating particles and making them collide is another way to produce particles. In Part II, we are going to analyse data coming from a scattering experiment.

1.3.2 Observables

One distinguishes two observables: the *decay length* and the *cross section*. Any other fundamental parameter, e.g. the mass of the particles, is then extracted from the comparison of predictions and measurements.

Decay length. The *decay length* is the first observable that was measured in particle physics. All unstable particles, fundamental as well as composite, have a different lifetime.

Cross section. The second main measurable quantity in HEP is the *cross section*. Classically, the cross section is the overlapping area of the projections of the target in the transverse plane and the projectile particle. A first generalisation was performed when studying diffraction in optics, the cross section being then defined in terms of intensities rather than areas [8]. The concept was further extended in particle physics: particles having no clear borders, the cross section cannot be properly defined as a physical area; eventually, it is interpreted as a *rate* of scattering. Techniques to compute cross sections in HEP will be discussed in Chap. 2. In this thesis, we are going to measure a cross section.

1.4 Challenges

Despite the remarkable precision achieved in HEP experiments, many questions remained unanswered; for instance:

- gravitation is not described;
- the asymmetry observed between matter and antimatter in the universe is not explained;
- evidence for dark matter and dark energy abound in the universe (more in Appendix 1.A);
- the mathematical structure of the SM is unexplained, as well as its nineteen input parameters (see Chap. 2);
- calculations from the SM are not always analytically feasible, resulting in difficulties to produce predictions (see Chap. 2).

In this thesis

We present the measurement of the cross section of the inclusive *b* jet production in proton-proton collisions with the CMS experiment. The goal of this analysis is to test our knowledge at the *TeV scale*.

The first part is dedicated to present the context of the measurement. First, some elements of theory are given in Chap. 2 in order to discuss the current understanding of HEP and of proton-proton collisions; the notions of cross section and jet will be more rigorously detailed. This will be followed by a description of the CMS experiment, our experimental set-up, in Chap. 3. In Chap. 4, the MC techniques, abundantly used for calculations in HEP, are discussed, and some models used in the second part are already discussed. A review of *b* physics closes the first part in Chap. 5.

The second part is dedicated to the measurement itself. First the strategy of the analysis will be described in Chap. 6. Then Chaps. 7–8 contain the analysis itself. The comparison of the measurement to predictions is presented in Chap. 9.

Finally, a third part is composed of prospects and of various appendices.

1.A Dark Matter and Dark Energy

As we already stressed, gravitation is too weak to compete with other interactions, and is therefore not described in HEP. But the existence of dark matter and dark energy is a strong motivation for BSM.

1.A.1 Dark Matter

Several observations suggest more matter to be in the universe than the radiated light may let it believe, i.e. some type of matter that does not interact electromagnetically and that cannot be found on earth; this unknown matter is called *dark matter*.

Historically, the main technique to detect dark matter has been to compute the difference between the luminous mass and the dynamical mass. This can be done at different scales:

- star clusters [9],
- galaxy clusters [10]
- and galaxy dynamics [11].

More recently, *gravitational lenses* even allowed to map dark matter in the universe [12]; in addition, anisotropies in the *Cosmic Microwave Background* may be partly explained by the presence of dark matter [13, 14]. This list is not exhaustive, but these observations are pointing to an important missing piece of modern HEP.

1.A.2   Dark Energy

The *cosmological constant* is necessary to explain the observed expansion of the universe with the theory of general relativity [15]. Since it can be understood as a contribution to the energy, it is called *dark energy*. It would account for around two thirds of the content of the universe [16]; on the other hand, unlike baryonic and dark matter, it would fill the entire universe quite uniformly. Its nature is totally unknown; and its density is too low to be detected in experiments as of today.

References

1. Tanabashi M et al (2018) Review of particle physics. Phys Rev D 98(3):030001
2. The decay of a lambda particle in the 32 cm hydrogen bubble chamber (1960). https://cds.cern.ch/record/39474
3. Perrin J (1930) Les atomes. Librairie Félix Alcan, Paris
4. Rutherford E (1914) LVII The structure of the atom. Philos Mag 27(159):488–498. https://doi.org/10.1080/14786440308635117. http://dx.doi.org/10.1080/14786440308635117. http://dx.doi.org/10.1080/14786440308635117
5. Kurzon (2017) Geiger–Marsden experiment expectation and result. Accessed: 2017-09-10. https://upload.wikimedia.org/wikipedia/commons/f/f9/Geiger-Marsden_experiment_expectation_and_result.svg
6. DESY A picture of the proton. http://www.desy.de
7. Abramowicz H et al (2015) Combination of measurements of inclusivedeep inelastic $e \pm p$ scattering cross sections and QCD analysis of HERA data. Eur. Phys. J. C75(12):580. https://doi.org/10.1140/epjc/s10052-015-3710-4. arXiv:1506.06042 [hep-ex]
8. Barone V, Predazzi E (2002) High-energy particle diffraction, vol. 565. Texts and monographs in physics. Springer, Berlin. isbn: 3540421076. http://www-spires.fnal.gov/spires/find/books/www?cl=QC794.6.C6B37::2002
9. Kapteyn JC (1922) First attempt at a theory of the arrangement and motion of the sidereal system. Astrophys J 55:302–328. https://doi.org/10.1086/142670
10. Zwicky F (1933) Die Rotverschiebung von extragalaktischen Nebeln. Helv Phys Acta 6:110–127 [Gen Relativ Gravit 41(207) (2009)]. https://doi.org/10.1007/s10714-008-0707-4
11. Rubin VC, Ford Jr WK (1970) Rotation of the Andromeda nebula from a spectroscopic survey of emission regions. Astrophys J 159:379–403. https://doi.org/10.1086/150317
12. Natarajan P et al (2017) Mapping substructure in the HST frontier fields cluster lenses and in cosmological simulations. In: Mon Not R Astron Soc 468(2):1962–1980. https://doi.org/10.1093/mnras/stw3385. arXiv:1702.04348 [astro-ph.GA]

13. Ade PAR et al (2016) Planck 2015 results. XIII. Cosmological parameters. In: Astron. Astrophys. 594:A13. https://doi.org/10.1051/0004-6361/201525830. arXiv:1502.01589 [astroph.CO]
14. Hu W (2017) Intermediate guide to the acoustic peaks and polarization. http://background. uchicago.edu/~whu/intermediate/intermediate.html. Accessed: 2017-10-21
15. Huterer D, Turner MS (1999) Prospects for probing the dark energy via supernova distance measurements. Phys Rev D60:081301. https://doi.org/10.1103/PhysRevD.60.081301. arXiv:astro-ph/9808133 [astro-ph]
16. Frieman J, Turner M, Huterer D (2008) Dark energy and the accelerating universe. Ann Rev Astron Astrophys 46:385–432. https://doi.org/10.1146/annurev.astro.46.060407.145243. arXiv:0803.0982 [astro-ph]

Chapter 2
Elements of Theory

Hypotheses non fingo
— Isaac NEWTON [1]

IN THIS CHAPTER, SOME IDEAS OF THE THEORETICAL ASPECTS of modern HEP
are presented, with a special emphasis on the topics underlying the measurement
presented in Part II: first the Standard Model (SM), and in particular Quantum Chromodynamics (QCD); then additional phenomenological models used in the treatment
of proton–proton collisions are introduced, in particular the *evolution equations*.

The approach given here does not reproduce the sequence of the historical discoveries: as an alternative of the two first sections, the development of fundamental
physics from the foundations to early days may be found in the appendix of this
chapter. The physics of the b quark will be the object of Chap. 5.

2.1 Introduction

The two main observables in HEP were already mentioned in Sect. 1.3.2: the *decay
rate* and the *cross section*. In both cases, phenomena with different numbers of
particles in the initial and final states have to be accounted for; the *amplitude of
transition* $\mathcal{M}$, or Matrix Element (ME), is written as follows [2]:

$$\mathcal{M} = \langle \text{final state} |\ \text{interaction Hamiltonian}\ |\text{initial state} \rangle \qquad (2.1)$$

The interaction Hamiltonian can be complicated—this will be discussed later in
the chapter. From now on, the discussion is restricted to the computation of the
cross section of any process $1 + 2 \rightarrow 3 + 4 + \ldots + N$, which can be represented
by the following diagram:

© Springer Nature Switzerland AG 2019

P. L. S. Connor, *Inclusive b Jet Production in Proton-Proton Collisions*, Springer Theses,
https://doi.org/10.1007/978-3-030-34383-5_2

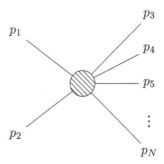

where the blob represents the interaction. The cross section can be deduced from the amplitude of transition by the following formula:

$$\sigma = \frac{S}{4\sqrt{(p_1 p_2)^2 - (m_1 m_2)^2}} \underbrace{\int \cdots \int}_{\substack{\text{phase} \\ \text{space}}} \overline{|\mathcal{M}|^2} (2\pi)^4 \delta^4 \left(p_1 + p_2 - \sum_{j=3}^{N} p_j \right) \quad (2.2)$$

$$\prod_{j=3}^{N} 2\pi\delta \left(p_j^2 - m_j^2 \right) \Theta(E_j) \frac{\mathrm{d}^4 p_j}{(2\pi)^4}$$

where

- the $p_i = (E_i, \mathbf{p}_i)$ stand for the four-momenta;
- the line over the squared ME stands for the summation (average) over the spins and over the colours in the initial (final) state if relevant,
- S is a combinatorial factor to avoid double counting when particles in the final state are identical;
- the integral runs over the phase space of all particles in the final state.
- the first Dirac delta function ensures that energy and momentum are conserved between initial and final states;
- the second Dirac delta function fixes the four-momenta of the outgoing particles to satisfy the mass condition, i.e. the outgoing particles have to be *on the mass shell*;
- the Heaviside function ensures that all particles in the final state have positive energy.

Among others, the Standard Model (SM) provides tools to compute the ME: this will be the topic of Sect. 2.2. In some cases however, such as at Large Hadron Collider (LHC), the possibilities of calculation within the SM will be limited, and one resorts to phenomenological models: this will be the topic of Sect. 2.3.

2.2 The Standard Model of High Energy Physics

The SM consists of a whole, self-consistent, auto-sufficient body of theory that aims at accounting for all high-energy phenomena that are unambivalently founded theoretically and strongly attested by several experiments.

In this section, we draw a portrait of the SM [3, 4]. First, a(n attempt of) definition of the SM is given (Sect. 2.2.1). Then, its Lagrangian formulation is outlined (Sect. 2.2.2), the couplings of the interactions are discussed (Sect. 2.2.3), and computations techniques with Feynman diagrams are introduced (Sect. 2.2.4). At the end of the section, the current difficulties of computation techniques within the SM are discussed (Sect. 2.2.5).

2.2.1 Definition

The SM is a *renormalisable* (Sect. 2.2.1.3) *relativistic quantum field theory* (Sect. 2.2.1.1) based on *local Gauge invariance* (Sect. 2.2.1.2) [5] with *19 input parameters* (Sect. 2.2.1.4), describing the electromagnetic, weak and strong interactions (already introduced in Chap. 1).

2.2.1.1 Quantum Field Theory

As a consequence from the marriage of quantic and relativistic physics, the number of particles cannot be conserved; requiring a fixed number of particles in a relativistic system would violate causality [4]. Therefore the fundamental objects of the SM are not particles but *quantum fields* [6]. Particles—what is eventually seen in a detector— are considered as excitation modes of these fields.

The quantum fields existing in the SM are summarised in Table 2.1a; fields may be classified according to different properties:

spin Fields in the SM can have spin 0 (scalar), $\frac{1}{2}$ (spinor) or 1 (vector). Spin-$\frac{1}{2}$ (spin-1) fields are usually considered as the fields of matter (interaction); they are listed in Table 2.1b (Table 2.1c). The BEH boson,[1] or simply Higgs boson, is the only spinless field.

transformation under boost Vector fields (spin-1), spinor fields (spin-$\frac{1}{2}$) and scalar fields (spin-0) undergo different relativistic transformations according to different laws [7, 8].

mass The mass plays also a rôle in the relation between the components of a spinor or vector field. For instance, neutrinos are massless and have only two components (Weyl spinor), while electrons are massive and have four components (Dirac spinor).

Fields have units of energy density.

---

[1]Named after the three physicists BROUT, ENGLERT and HIGGS.

Table 2.1 In the SM, the fields are classified according to their symmetries: the spin, the mass and the charge

spin	mass	type of field	d.o.f.	particles	diagram
0	yes	scalar	1	BEH scalar boson	- - - - - - - - - -
$\frac{1}{2}$	no	Weyl spinor	2	neutrinos	⟶
	yes	Dirac spinor	4	charged leptons	⟶
	yes	Dirac spinor	4	quarks	⟶
1	no	massless vector	2	photon	∿∿∿∿∿
	yes	massive vector	3	weak bosons	∿∿∿∿∿
	no	massless vector	2	gluons	⦿⦿⦿⦿⦿

(A) Fields in the Standard Model.

type	particles	electromagnetism	weak forces	strong forces
charged leptons	$e,\ \mu,\ \tau$	$Q = \pm 1$	yes	white
neutral leptons	$\nu_e,\ \nu_\mu,\ \nu_\tau$	$Q = 0$		
up quarks	$u,\ c,\ t$	$Q = \frac{2}{3}$	yes	RGB
down quarks	$d,\ d,\ b$	$Q = -\frac{1}{3}$		

(B) Particles of matter are all half-spin particles. Particles have been ordered by increasing mass. Six *flavours* exists for the quarks. Except for the neutrinos that are neutral, every elementary particle of matter has a matching antiparticle that has the same properties but an opposed electric charge, contained in the same field.

Gauge boson	symbol	interaction	effective range
photon	γ	electromagnetism	infinite
neutral weak boson	Z^0	weak interaction	nuclear scale
charged weak bosons	$W^\pm$		
gluons	g	colour interaction	nuclear scale

(C) Particles of interaction are all 1-spin particles. The eight gluons being perfectly symmetric in the colour space, they cannot be distinguished from one another and are referred altogether only once.

2.2.1.2 Local Gauge Invariance

Before describing it directly in the SM, it may be worthwhile to retrace the principle of Gauge invariance from earlier theories, in classical electrodynamics and in quantum mechanics.

In classical electrodynamics. Gauge invariance has already existed in classical electromagnetism; the electric and magnetic fields can be deduced from a scalar and a vectorial potentials ϕ and $\mathbf{A}$: $\mathbf{E} = -\nabla\phi - \frac{\partial \mathbf{A}}{\partial t}$, $\mathbf{B} = \nabla \times \mathbf{A}$. Any *Gauge transformation* Λ leads to the same evolution; in other words Maxwell equations are invariant under any transformation of the following form: $\phi \longrightarrow \phi + \frac{\partial \Lambda}{\partial t}$, $\mathbf{A} \longrightarrow \mathbf{A} + \nabla\Lambda$. In four-vectorial notation, the electromagnetic field is written $A = (\phi, \mathbf{A})$, hence the notation for the vector field for the photon. The Gauge invariance corresponds to the invariance of the equations of motion under *local transformations* of the fields, meaning transformations which are not uniform but can smoothly vary in space-time.

In quantum mechanics. Gauge invariance also applies in non-relativistic quantum mechanics, and is equivalent to a redefinition of the phase[2] of the wave function [9]; in this case, one can observe that this symmetry corresponds to the group $U(1)$. Several properties may be derived consequently to the property of Gauge invariance, such as the charge conservation and the conservation of the amplitude of probability in quantum mechanics.

In the Standard Model. In the SM, the Gauge invariance is generalised to more complex groups like $SU(2)$ ($SU(3)$) for weak (strong) interactions [5, 8, 10]; interactions derived from Gauge symmetries are called *Gauge interactions*. The locality of the Gauge interactions between two fields can be derived consequently to the locality of Gauge invariance; interactions at different places of space-time would indeed violate causality [4]. The vector fields, carrying the interaction in the SM, are thus said to be *Gauge fields*. Moreover, the Gauge symmetry is said to be internal, because it is a symmetry in the space of the charge (e.g. $U(1)$ for the electric charge, $SU(3)$ for the colour charge). In general, group theory (and Lie algebras) play a very important rôle in the description of symmetries of the SM [5, 11]; here, we only mention some properties. The cases of Quantum Electrodynamics (QED) and of Quantum Chromodynamics (QCD) are taken, the former for its simplicity, the second for its relevance in this thesis:

QED   Electrically charged fields ψ interact by exchanges of photons A. The dynamics is invariant under the following transformation:

$$\text{electron}\quad \psi \rightarrow e^{ig\alpha}\psi \tag{2.3}$$

$$\text{photon}\quad A_\mu \rightarrow A_\mu - \frac{1}{g}\partial_\mu\alpha \tag{2.4}$$

The $U(1)$ Gauge invariance implies several important properties of QED: the conservation of the charge, the null mass of the photon, or the absence of self-interactions of the photon. The electric charge can take two values: $\pm e$, exactly like in classical electromagnetism.

QCD   Coloured fields ψ interact by exchanges of gluons A. One counts three colours (anti-colours) for quarks (anti-quarks)[3] and eight colours for gluons. The strong interaction relying on $SU(3)$, the Gauge transformation takes a more complicated form than in QED:

---

[2]The word *Gauge* is indeed an old word for "phase".

[3]The term of colour is therefore taken in analogy to the three primary colours and explain the *chromo* in Quantum Chromodynamics.

$$\text{quark} \quad \psi \rightarrow e^{ig_s \frac{\lambda^a}{2}\theta^a} \psi \tag{2.5}$$

$$\text{gluon} \quad A_\mu^a \rightarrow A_\mu^a - \frac{1}{g_s}\partial_\mu\theta^a + f^{abc}\frac{\lambda^b}{2}A_\mu^c \tag{2.6}$$

with the Gell-mann matrices λ^a ($a = 1, \ldots, 8$) and f^{abc} the structure constants of $SU(3)$. The existence of eight colour states for the gluon is directly related to the structure of $SU(3)$ [2]. Unlike QED, this additional term in QCD will be responsible for self-interactions of the gluon field.

The different symmetries corresponding to the charges of the three fundamental interactions in the SM are described by different *unitary groups*, with *group structure* $U(1) \times SU(2) \times SU(3)$. The reason for this structure is an open question in HEP.

2.2.1.3 Renormalisation

Most Quantum Field Theorys (QFTs) do not lead to finite amplitudes of transitions. Only a limited number of interactions can be considered without rendering the theory ill-defined, with unresolvable *ultraviolet divergences*. Fortunately, there exist QFTs on which a procedure of *renormalisation* can be applied at the cost of

- introducing a (non-physical) *renormalisation scale*,
- renormalising the field,
- and redefining the couplings and the masses.

The SM is indeed renormalisable [2, 12, 13]. Some aspects related to renormalisation issues will be discussed later on.

2.2.1.4 Parameters

The SM requires 19 parameters, summarised in Table 2.2, which are not constrained by the theory and need to be determined experimentally:

- masses of the charged leptons,
- masses of the quarks,
- fundamental parameters,
- CKM matrix,[4]
- extra parameters.

While the three charged leptons are massive, the neutrinos are massless in the SM. Experimentally though, it has been measured that the neutrinos are also massive, but since their masses are very small and since some doubts on their exact field properties still remain, their masses are not included in the SM yet.

---

[4]Named after the three physicists CABIBBO, KOBAYASHI and MASKAWA.

Table 2.2 The nineteen parameters of the SM have been measured or constrained experimentally. The mass of the leptons, the mixing angles, the CP violation phase and the extra parameters are absolute, whereas the mass of the quarks and the fundamental parameters, which can be obtained from the Gauge couplings, depend on the renormalisation scheme (especially, the fundamental parameters are here given at the scale $Q^2 = M_Z^2$). The values from [15]

	Description	Symbol	Value
Charged lepton masses	Electron	m_e	511 keV
	muon	m_μ	105.7 MeV
	tauon	m_τ	1.78 GeV
Quark masses	Up	m_u	~2 MeV
	Down	m_d	~4.5 MeV
	Strange	m_s	~87 MeV
	Charm	m_c	~1.3 GeV
	Bottom	m_b	~4.2 GeV
	Top	m_t	~173 GeV
Fundamental constants	Fine structure constant	α	$\frac{1}{128.957 \pm 0.020}$
	Weinberg angle	$\sin^2 \theta_W$	0.23116 ± 0.00012
	Strong coupling	α_S	0.1184 ± 0.0007
CKM matrix	12-mixing angle	θ_{12}	13.1°
	23-mixing angle	θ_{23}	2.4°
	13-mixing angle	θ_{31}	0.2°
	CP violation phase	δ	0.995
Extra parameters	QCD vacuum angle	Θ_{QCD}	~0
	Vacuum expectation value	v	246 GeV
	Higgs mass	m_H	125.09 ± 0.24 GeV

The mass of the quarks is affected by the renormalisation. The masses of Heavy Flavour (HF) quarks (corresponding to quarks with mass above 1 GeV, i.e. charm, bottom and top) are given at a scale of the order of the physical mass.

The CKM matrix is involved in the mixing of the quark flavours, necessary for the description of the weak decays of massive quarks into lighter quarks of different charge. The phase parameter δ describes the *weak CP violation* [14] (not discussed here).

The fundamental parameters describe the intrinsic strength of an interaction, in contrast to the (electric, weak, or strong) charges that describe the behaviour of the particle according to this interaction. The fundamental parameters can be directly related to the Gauge couplings g, which appear explicitly in the Gauge transformation. Their values also depend on the renormalisation scale; in particular, the evolution of their values will be discussed in more detail in Sect. 2.2.3.

The Θ_{QCD} parameter is related to the *strong CP violation*, unobserved but predicted by the SM (not discussed here).

Finally, the *vacuum expectation value* and the *Higgs mass* are related to Higgs physics (not discussed in this thesis).

2.2.2 Lagrangian

We describe the mathematical expression of the SM to compute predictions.

The commonest form in which the SM is expressed consists in a Lagrangian density $\mathcal{L}_{SM}$. In principle, one can apply the Principle of Least Action to deduce the Equations of Motion (EOM):

$$S = \int \mathcal{L}_{SM}\left(\phi, \partial_\mu \phi\right) \, \mathrm{d}^4 x \quad \longrightarrow \quad \text{EOM} \equiv \delta S = 0 \tag{2.7}$$

The amplitude of transition can then be determined from the action ($\mathcal{M} \sim \exp(iS)$). In practice however, because of the difficulties to solve them, the EOMs are rarely directly used in analyses such as the one presented in this thesis. But an interpretation may be read directly from each term of the Lagrangian, and generic rules how to compute cross sections have been invented by Richard FEYNMAN; before going to this topic (Sect. 2.2.4), the Lagrangian is further described.

The Lagrangian contains a large number of terms (varying according to the representation), and can be divided in certain sectors:

$$\mathcal{L}_{SM} = \mathcal{L}_{EW} + \mathcal{L}_{QCD} + \mathcal{L}_{Higgs} + \mathcal{L}_{Yukawa} \tag{2.8}$$

2.2.2.1 Electroweak Sector

The electromagnetism and weak interactions are entangled in the same sector, known as Electroweak (EW) [3]. The involved fermions are *leptons* (either charged leptons or neutrinos) and *quarks*, interacting electromagnetically by exchanges of photons or weakly by exchanges of $W^\pm$ or Z^0 bosons. The photon has infinite range and is massless, while the weak bosons exist only at the scale of the nucleus and are massive. Each field possesses excitation states corresponding to particles and to antiparticles, according to the electric charge. In addition, fermions exist in three generations, as shown in Table 2.1b:

charged leptons	electron (e), muon (μ), tauon (τ)
neutral leptons	partner neutrinos (ν_e, ν_μ, ν_τ)
up-type quarks	up (u), charm (c), top (t)
down-type quarks	down (d), strange (s), bottom (b)

The discussion is now restricted to electromagnetism, i.e. to QED, leaving out the weak interaction. A common representation of the Lagrangian of QED reads as follows:

$$\mathcal{L}_{QED} = \sum_{fermions} \bar{\psi}_f \left(i\slashed{D} - m\right) \psi_f - \frac{1}{4} F_{\mu\nu} F^{\mu\nu} \tag{2.9}$$

where the electromagnetic field (i.e. the photon) is hidden both in the *covariant derivative D* (the slash only indicating how the components of the spinor are combined) and in the *electromagnetic tensor* F:

$$D_\mu = \partial_\mu - i A_\mu \tag{2.10}$$
$$F_{\mu\nu} = \partial_\mu A_\nu - \partial_\nu A_\mu \tag{2.11}$$

After reorganising the terms, one can distinguish three types of terms:

mass terms   order two in the fields (e.g. $-\frac{1}{2} A_\mu A^\mu$)
interaction terms   order three or four in the fields (e.g. $e^2 \bar\psi \gamma^\mu \psi A_\mu$)
kinematic terms   terms involving derivatives (e.g. $i\bar\psi\gamma^\mu \partial_\mu \psi$)

2.2.2.2 Quantum Chromodynamics Sector

The Quantum Chromodynamics describes the strong nuclear interactions. Quarks and antiquarks interact by exchanges of gluons [16]. The $SU(3)$ symmetry implies that quark (antiquarks) exist in three colour (anti-colour) states, and gluons in eight; the symmetry is however perfect and the states cannot be distinguished. The range of this interaction is that of the nuclear scale. QCD is not directly sensitive to the flavour; the Lagrangian of QCD is invariant with respect to the flavour, since the different quarks have different masses, the *flavour democracy*, or *flavour blindness*, only takes place at energies where all the quark masses are negligible.

The Lagrangian of QCD is similar to the Lagrangian[5] of QED (Eq. 2.9):

$$\mathcal{L}_{\text{QCD}} = \sum_{\text{quarks}} \bar\psi_q^i \left(i \slashed{D}^{ij} - m\delta^{ij} \right) \psi_q^j - \frac{1}{4} G_{\mu\nu}^a G^{a\mu\nu} \tag{2.12}$$

where the strong field (i.e. the gluon) is hidden both in D and G:

$$D_\mu^{ij} = \delta_{ij}\partial_\mu - i g_s \frac{\lambda_a^{ij}}{2} B_\mu^a \tag{2.13}$$
$$G_{\mu\nu}^a = \partial_\mu B_\nu^a - \partial_\nu B_\mu^a + g_s f^{abc} B_\mu^b B_\nu^c \tag{2.14}$$

In comparison with the Lagrangian of QED in Eq. 2.9, an additional Latin index runs on the colour ($i, j = 1, 2, 3$ for quarks, $a, b, c = 1, \ldots, 8$ for gluons). Similar terms are found, with additional interaction terms for the gluon field with itself.

2.2.2.3 Yukawa and Higgs Sectors

The Higgs boson is the most recently discovered particle in the SM [17, 18]. It is involved in a complex procedure of symmetry breaking of the electroweak sector, allowing the presence of mass terms in the Lagrangian for the fermions (Yukawa

---

[5]The term related to Θ_{QCD} is here neglected, since there is no experimental evidence for it.

sector) and for the weak bosons (Higgs sector) without violating the Gauge symmetry in the EW sector[6] [19, 20]. The Higgs sector also describes the self-interactions of the Higgs boson.

2.2.3 Running Couplings

The QFT of an interaction is characterised by a coupling α, directly related to the constant g in the Gauge transformation (e.g. in Eqs. 2.3 and 2.5). Gauge couplings have already been briefly mentioned while discussing the nineteen input parameters of the SM (Sect. 2.2.1.4). For instance, the electromagnetic coupling is the *fine structure constant*.

Gauge couplings are crucial because they determine the perturbative or non-perturbative nature of the interaction. Different Gauge structures are related to different behaviours of the respective couplings; especially, while the QED coupling *decreases* from small to large distances, the QCD coupling *increases*. This property is related to the self-interactions of the gluon field.

In QED and QCD, this coupling is said to be a *running coupling*, since it depends on the energy scale Q. In a first approximation, QED and QCD are scale invariant, which means that the interaction will be the same at all scale. The *scale violation* is only logarithmic [21], but when different orders of magnitudes are involved, the variations of the couplings cannot be neglected.

In addition, the couplings of QED and QCD have qualitatively opposite behaviours:

QED   The coupling α, hence the effect of the electric charge, decreases with larger distances. However, at very high energies, i.e. at very small distances, the coupling will become closer to unity; the validity of the perturbative regime is expected to break down at some point. While in atomic physics, the fine structure constant[7] is close to $1/137$, in HEP at LHC, at around $100\,\text{GeV}$, it is close to $1/127$; the non-perturbative regime is still far from the accessible phase space.

QCD   The behaviour is opposite: the coupling α_S increases at small energies, *i.e* at large distances; this is called the *asymptotic freedom*. At small energies, the QCD is in the *non-perturbative regime*. In QCD, only the ME at high energy can potentially be calculated; hadrons can rarely be described with perturbative equations. One can write the strong running coupling as follows [22][8]:

$$\alpha_S(Q^2) \propto \frac{1}{\ln\left(\frac{Q^2}{\Lambda_{QCD}^2}\right)} \tag{2.15}$$

---

[6]Unlike the EW sector, the QCD sector involves vectors; mass terms are therefore not forbidden by the Gauge symmetry in QCD.

[7]At low energy, it is given by $\alpha = \frac{e^2}{4\pi\epsilon_0 \hbar c}$ and measured with a very high precision to $7.297, 352, 5698(24) \times 10^{-3}$ [15].

[8]This corresponds to the renormalisation scheme called *minimal subtraction*.

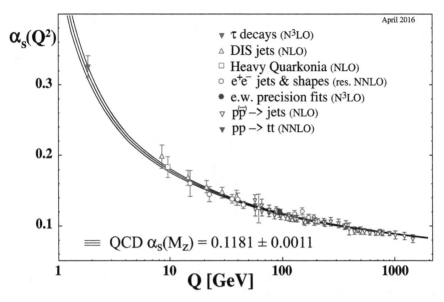

Fig. 2.1 Summary of measurements of α_S as a function of the energy scale Q. The respective degree of QCD perturbation theory used in the extraction of α_S is indicated in brackets (…). Figure reproduced with permission of authors [15]

where $\Lambda_{QCD} \sim 200$ MeV, typically the scale of hadronic masses; the transition between the perturbative and non-perturbative regimes is given by the Λ_{QCD} constant (below Λ_{QCD}, Eq. 2.15 has no validity). The variation, or *running*, of α_S is significant, as shown in Fig. 2.1.

Measured values of the couplings have already been given in Table 2.2; they correspond to *renormalised* couplings. By convention, the *couplings'* values are given at the peak of mass of the Z^0 boson $Q^2 = M_Z^2 \approx (91.2\text{GeV})^2$, i.e. well in the perturbative regime. While α_S is the direct analog to α for the strong force, θ_W, the *Weinberg angle*, is of different nature, as it is involved in the mixing of electromagnetism and weak force within the electroweak theory. A recent review of the latest measurements of the strong coupling may be found in [23].

2.2.4 Calculations with Feynman Diagrams in the Perturbative Regime

Richard FEYNMAN invented a graphical representation of the terms appearing in the Lagrangian to treat calculations in the perturbative regime, called *Feynman diagrams* [24].

These diagrams can be seen as representations of the dynamics, how the fields interact, and allow to understand the calculations. Although not used explicitly to perform calculations in this thesis, it is worthwhile giving some first principles in order to discuss theoretical predictions.

A Feynman diagram connects the particles in the initial (on the left hand side) and final states (on the right hand side):

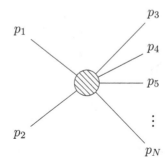

The time (space) coordinate is represented horizontally (vertically). What happens in the blob is represented by *lines* and *vertices*:

lines   The lines correspond to the type of field (see Table 2.1a). Each one corresponds to a factor in the calculation, called *propagator*.

vertex   Interactions are represented by joint lines:

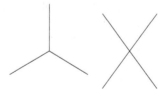

The four-momentum has to be conserved at every vertex.

All particles inside (outside) of the blob are said to be *virtual* (*real*). A real process being characterised only by its initial and final states, as described in Eq. 2.1; virtual particles cannot be observed. A virtual particle may even have a different mass than its corresponding real manifestation.

In principle, the interaction Hamiltonian will be made of *all* the possible combinations of vertices respecting the laws of conservations between the initial and final states. The exact *Feynman rules*, translating the diagrams into terms entering the computation of the ME, may be found for instance in Ref. [4]; here, we only give the principles.

From the Lagrangian, one can deduce all possible diagrams in QED and QCD:

QED   Only one type of vertex exists: the interaction of a fermion with a photon:

$\bar{f}$

f

γ

QCD The analog diagram to the QED one exists:

$\bar{f}$

g

f

In addition, since the gluon can interact with itself, 3- and 4-leg vertices are possible:

Real processes however always involve two scattering particles in the initial state and at least two product particles in the final state: 3-leg vertices do not correspond to any real process and must be combined at least in pairs; 4-leg vertices may however take place without being combined.

2.2.4.1 Example

In principle, in order to compute the ME of a given process (see Eq. 2.1), one should consider *all* possible combinations of the interactions. To give a preview of the complexity, one can take the example of $e^+e^- \to \mu^+\mu^-$ at an energy scale $\mathcal{O}(1\,\text{GeV})$, where the situation is as simple as possible since only QED is involved with distinct initial and final states.

$$i\mathcal{M}_{\mu^+\mu^-} = \quad \text{(tree-level)} \quad (2.16)$$

$$+ \quad \text{} \qquad \text{(1-loop diagrams)} \qquad (2.17)$$

$$+ \quad \text{} \qquad\qquad (2.18)$$

$$+ \quad \text{one similar diagram} \qquad\qquad (2.19)$$

$$+ \quad \text{} \qquad\qquad (2.20)$$

$$+ \quad \text{one similar diagram} \qquad\qquad (2.21)$$

$$+ \quad \text{} \qquad\qquad (2.22)$$

$$+ \quad \text{three similar diagrams} \qquad\qquad (2.23)$$

$$+ \quad \dots \qquad \text{(higher loop multiplicity)} \qquad (2.24)$$

Including more complicated diagrams means computing higher-order terms in the calculation, but they are rarely considered. The given diagram is one of the simplest ones; in general, complications arise very soon:

– A more realistic view should be to consider additional radiations:

Such a radiation could take place before the annihilation (Initial-State Radiation, or ISR), or after the decay of the virtual photon (Final-State Radiation, or FSR). These additional radiations are important: *soft* (i.e. at low energy) and *collinear* (i.e. collimated with the emitter) radiations cause the ME calculation to diverge – this is called *infrared divergences*, unless *loop diagrams* (such as Eq. 2.22) are included.

– When describing $e^+e^- \to e^+e^-$, two diagrams (instead of only one) should be considered already at tree level:

From left to right, these two diagrams are respectively said to be in the s-channel and t-channel (see also Appendix 2.B.2). The multiplicity of diagrams can become extremely high as soon as one includes higher orders to the calculation, or in other words, more loops and more legs.

– At higher energies, one should also start considering diagrams involving weak bosons $W^\pm$ and Z^0, implying additional loop diagrams. Moreover, QCD loops should also be considered, adding yet other diagrams.

As a conclusion of this example, the representation in Feynman diagrams allows to see how even the most elementary processes may require advanced calculations.

2.2.4.2 Vocabulary

We introduce now some common terms employed to characterise calculations. The context is fixed to QCD.

Inclusive and exclusive cross sections. Given a certain process $ij \rightarrow F$, the *inclusive* (*exclusive*) cross section corresponds to the final state including (excluding) all extra radiations. The inclusive cross section can be seen as a sum of exclusive cross sections with k extra radiations in the final state:

$$\hat{\sigma}_{ij \rightarrow F}^{\text{incl}} = \sum_{k} \hat{\sigma}_{ij \rightarrow F+k}^{\text{excl}} \tag{2.25}$$

Leading Order and Next-to-Leading Order. The inclusive cross section of $ij \rightarrow F$ can be further decomposed in MEs, the diagrams with identical final state correspond to loop diagrams:

$$\hat{\sigma}_F^{\text{incl}} = \sum_{k} \int d\Phi_{F+k} \sum_{l} \left| \mathcal{M}_{F+k}^{(l)} \right|^2 \tag{2.26}$$

where l stands for the number of loops in the diagram ($d\Phi$ is a common short notation of the whole element of phase space). Then, the terms may be rearranged in orders of the strong coupling:

$$\hat{\sigma}_F^{\text{incl}} = \sum_{n} \hat{\sigma}_F^{N^n LO} \tag{2.27}$$

where n stands for the order in the strong coupling. Each leg (loop) contributes with one (two) order(s) in the coupling. One commonly defines:

$n = 0$ Leading Order
$n = 1$ Next to Leading Order
$n = 2$ Next to Next to Leading Order

(above, one simply writes N^nLO, but higher orders are rarely considered). The number of terms contributing to a cross section at the first orders is shown in Table 2.3. The *Born cross section* designates LO for $2 \rightarrow 2$ processes, i.e. in the most elementary final state. Finally, the *tree level* corresponds to diagrams without loops, regardless of the number of legs.

Virtuality. The conservation of the momentum only applies to the legs of a diagram; internal lines do not respect this constraint. Let p be the four-momentum of an internal line, p^2 is its *virtuality*. If $p^2 \approx m^2$ ($p^2 \neq m^2$) where m is the mass of the corresponding particle, then the particle is said to be *on-shell* (*off-shell*). In diagrams, off-shell particles are sometimes noted with a star (e.g. γ^* stands for a virtual photon).

2.2.5   Current Limitations of Calculations within the SM

The SM is a very powerful theory and has been verified to an extreme precision; the nineteen input parameters have been measured (or at least constrained) with very good precision.

Table 2.3 Number of loops and legs contributing to the different orders in the strong coupling

Additional terms	First factor		Second factor	
	# loops	# legs	# loops	# legs
LO	0	0	0	0
NLO	1	0	0	0
	0	1	0	1
NNLO	2	0	0	0
	1	0	1	0
	1	1	0	1
	0	2	0	2
etc.	$\vdots$	$\vdots$	$\vdots$	$\vdots$

However, it still suffers from certain major difficulties that we encounter in proton–proton collisions: the treatment of non-perturbative QCD and the complexity of the perturbative calculations due to the high multiplicity of the final states. This will motivate the phenomenological approach presented in the next section of this chapter.

2.2.5.1 Non-perturbative QCD

In QCD, low-energy systems are characterised by the property of *confinement*: quarks and gluons never appear alone and free, but always in bound states, i.e. in *hadrons*[9] [26, 27]. The study of the scattering of partons requires necessarily the scattering of hadrons; it is therefore essential to understand and to be able to describe these objects although they cannot be described with perturbation theory.[10]

2.2.5.2 High-Multiplicity Final States

In order to be able to apply perturbative QCD (pQCD), one needs to reach the regime of asymptotic freedom. This can only be achieved at energies $Q^2 \gg \Lambda_{QCD}^2$, which also means $Q^2 \gg m_{proton}^2$; in LHC conditions, the phase space is large enough to allow the production of hundreds of additional particles.

Unfortunately the calculations of cross sections (Eq. 2.2) with high multiplicity in the final state are most of the time too complex to be solved completely analytically:

---

[9]One should mention an extremely dense state of matter where partons are not confined in hadrons but deconfined in a very strong colour field: the *quark-gluon plasma* [25]. This kind of state is studied by the ALICE experiment at LHC.

[10]Although we are presently not concerned with it, one should at least mention another, non-phenomenological approach of non-perturbative QCD: *lattice QCD* [28, 29]. In this approach, QCD systems are discretised and treated numerically with very large computation resources. Good progresses are being achieved, but lattice QCD is currently not used in the present context.

Table 2.4 Number of diagrams in e^+e^- collisions for final states with a pair $u\bar{u}$ and extra gluon radiations at the tree level [30]

Final state	# diagrams
$u\bar{u}g$	2
$u\bar{u}gg$	8
$u\bar{u}ggg$	50
$u\bar{u}gggg$	428
$u\bar{u}ggggg$	4670

as soon as more than a few particles are present in the final state with several orders of accuracy in the perturbative expansion. For instance, the number of diagrams at tree level for $e^+e^- \rightarrow u\bar{u}X$ (where X stands for additional gluon radiations) is shown in Table 2.4. Most of them require Monte Carlo (MC) techniques (introduced in Chap. 4); the different theoretical predictions will differ on the order of the calculation, and on the phenomenological treatment of the additional radiations (described in Sect. 2.3).

2.3 Phenomenology

According to the type of collisions under study different phenomenological approaches exist. For *elastic scattering* (both protons remain intact) and *diffractive dissociation* (at least one proton gets destroyed), a common approach to treat such collisions is the Regge theory, where the interaction is described in terms of exchanges of a *pomeron*[11] [33, 34].

In this thesis, we shall rather consider inelastic non-diffractive scattering at LHC, where large exchanges of momentum take place. Indeed, in this context, the property of *factorisation* can be applied in order to extract and treat part of the problem in the perturbative regime (Sect. 2.3.1). Starting from this property, we describe the phenomenology of proton–proton scattering, first with hadrons and jets (Sect. 2.3.2), then with an overview of a scattering itself (Sect. 2.3.3). Finally, we close the chapter with a discussion of the different types of factorisations (Sect. 2.3.4).

2.3.1 Factorisation

An important property allows to separate the hadronic cross section into contributions from the interaction at parton level and from the hadron structure; this property is known as *factorisation*, and is expected to apply for the following processes [35]:

---

[11] As part of my contributions to the CMS collaboration, I also had the chance to participate to two measurements of the minimum-bias hadronic production in proton–proton collisions, where the different contributions of the total cross section are measured in detail [31, 32].

- in Deeply Inelastic Scattering (DIS):

$$l + h \rightarrow l' + X \tag{2.28}$$

where h (l, X) stands for any hadron (lepton, anything);
- in electron-positron scattering:

$$e^+ + e^- \rightarrow h + X \tag{2.29}$$

- the Drell-Yan (DY) process:

$$h + h' \rightarrow \mu^+ + \mu^- + X \tag{2.30}$$
$$h + h' \rightarrow e^+ + e^- + X \tag{2.31}$$
$$h + h' \rightarrow W + X \tag{2.32}$$
$$h + h' \rightarrow Z + X \tag{2.33}$$

- jet production:

$$h + h' \rightarrow j + X \tag{2.34}$$

where j stands for jet (see Sect. 2.3.2.2);
- heavy quark production:

$$h + h' \rightarrow Q + X \tag{2.35}$$

where Q stands for heavy quark.

Theoretically, it has been proven only for the three first processes for *leading twist*.[12] Experimentally however, the factorisation is believed to hold in presence of a *hard* process, i.e. at scale $\mu_F^2 \gg \Lambda_{QCD}^2$, corresponding to the regime of asymptotic freedom; but the fact that the same Parton Distribution Functions (PDFs) work for different types of scatterings (typically ep and pp collisions) is a very strong argument in favour of the legitimacy of the factorisation.

Technically, a *factorisation scale* μ_F^2 must be chosen: below this scale, the contributions to the hadronic scattering are contained in the PDFs; above this scale, the process is described by the partonic scattering (described in Sect. 2.2). Since this scale μ_F^2 is not physical, a PDF does not represent any fundamental quantity: it is only a way of separating (or factoring out, hence the name) the non-perturbative regime from the whole process; however, this separation also is not exact, and part of the perturbative regime is also *de facto* included in the PDF. As of today, PDFs cannot be computed and must be extracted (or fitted) from measurements. Different PDF extractions, or *sets*, are available, according to the data used to perform the extraction or to

---

[12]The twist corresponds to the mass dimension minus the spin; the cross section can be expanded in orders of the twist; in the factorisation, correction terms in $\ln\left(Q^2/\Lambda_{QCD}^2\right)^{m<2n}/Q^{2n}$ are neglected.

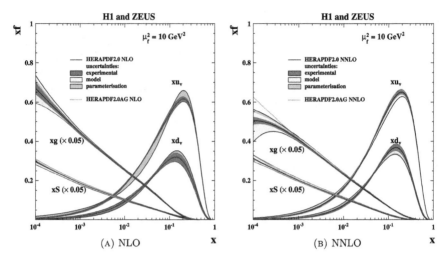

Fig. 2.2 Extraction of the PDFs of the proton at the HERA collider in *ep* scattering. The four curves correspond to up and down valence quarks, to gluons and to sea quarks. The PDFs are extracted in different sets according to the order. Both cases are qualitatively similar: the up and down valence quarks show a bump at high *x*, while the sea quarks and the gluons become dominant at low *x*. Figure reproduced with permission of authors [43]

the type of fit; for instance (alphabetically ordered, non-exhaustive list): ABM [36], CTEQ [37], HERAPDF [38], MMHT [39], MSTW [40] or NNPDF [41, 42].

Different schemes exist: we shall here consider the *collinear factorisation* with DGLAP[13] evolution. In this paradigm, PDFs are noted $f_{i/h}(x, \mu_F^2)$, where x_i corresponds to the *momentum fraction* of the hadron momentum carried by the parton i. (Other evolution and factorisation schemes will be discussed later in Sect. 2.3.4.)

The factorisation was initially introduced in the context of DIS:

$$\sigma_{lh \to F}^{incl} = \underbrace{\sum_{i \in g, q, \bar{q}} \int_0^1 dx_i \, f_{i \in h}(x_i, \mu_F^2)}_{\text{extraction from PDF}} \times \underbrace{\hat{\sigma}_{li \to F}^{incl}(x_i, \mu_F^2)}_{\text{parton-level process}} \qquad (2.36)$$

PDFs are universal in the sense that they should be common to all processes to which factorisation applies; however, different choices of PDFs apply according to the order of the ME. As an illustration, in Fig. 2.2 are shown the extractions at NLO and NNLO performed at the HERA collider (already introduced in Sect. 1.2.4) with the combined data from the H1 and ZEUS experiments [43].

Since they are universal, the same PDFs can in principle be used in the factorisation for hadron-hadron scattering as follows:

---

[13]Named after the five physicists DOKSHITZER, GRIBOV, LIPATOV, ALTARELLI and PARISI.

$$\sigma_{h_1 h_2 \to F}^{\text{incl}} = \sum_{i \in g, q, \bar{q}} \int_0^1 \mathrm{d}x_i \, f_{i \in h_1}(x_i, \mu_F^2) \times \tag{2.37}$$

$$\times \sum_{j \in g, q, \bar{q}} \int_0^1 \mathrm{d}x_j \, f_{j \in h_1}(x_j, \mu_F^2) \times \tag{2.38}$$

$$\times \hat{\sigma}_{ij \to F}^{\text{incl}}(x_i, x_j, \mu_F^2) \tag{2.39}$$

The hadron-hadron cross section may also be seen, at a given factorisation scale μ_F^2, as the sum on the gluons and the flavours of quarks of the corresponding parton-parton cross sections weighted by the PDFs.

Here, we shall focus on proton–proton interactions in the context of LHC, such as the ones that will be analysed in Part II. The application of factorisation will be described later on.

2.3.2 Objects in the Initial and Final States

Before detailing the interaction of the proton–proton cross section, it is necessary to define what can be found in the initial and final states of such a scattering. As already mentioned in Sect. 2.2.5.1, partons are never found free but always clustered into hadrons. Hadrons themselves are usually produced in the form of *jets*, corresponding qualitatively to the production coming from a parton in the final state of a strong interaction.

2.3.2.1 Hadrons

Hadrons are composite, colour-neutral, bound states holding together as a result of confinement property of the strong interaction.

Two kinds of hadrons exist, according to their *valence quarks*: just as the valence electrons that determine the properties of chemical species, the valence quarks determine the properties of the hadrons[14]:

baryons   which are triplets of quarks or antiquarks,
mesons   which are pairs of quark-antiquark.

Hadrons are also characterised by their lifetime. The proton is the only stable hadron,[15] which is the reason for which, in practice, hadronic collisions are mostly

---

[14] *Hadron* means *strong* in Old Greek, in relation to the strong force. *Baryon* (*mesons*) means *heavy* (*middle*). However, *leptons*, for *light*, are no hadrons. The etymology is purely based on empirical point of view.

[15] The neutron may be stable only if it is bound into a nucleus; this having been said, its lifetime being around fifteen minutes, it may also be considered as stable in the decay products of a collision.

Table 2.5 Selection of particles typically found in the final state in the detector with decay lengths [15]. The value are given without uncertainty (hence the $\sim$ symbol) to give a general survey of the behaviour of the hadrons in a typical modern experiment. Unless relevant, the antiparticle is not specified

Particle	Symbol	Content (if composite)	$c\tau$
Electron	e	–	∞
muon	μ	–	$\sim$659m
tauon	τ	–	$\sim$87 μm
Proton	p	uud	∞
Neutron	n	udd	2.64km
Charged pion	$\pi^\pm$	$\bar{u}d$	$\sim$7.80m
Neutral pion	π^0	$\bar{u}u$	$\sim$25.5nm
Charged kaon	$K^\pm$	$\bar{u}s/\bar{s}u$	$\sim$3.71m
Neutral short-lived kaon	K^0_S	$\bar{d}s/\bar{s}d$	$\sim$2.68cm
Neutral long-lived kaon	K^0_L	$\bar{d}s/\bar{s}d$	$\sim$15.4m
Positive sigma	Σ^+	uus	$\sim$2.40cm
Negative sigma	Σ^-	dds	$\sim$4.43cm
Neutral xi	Ξ^0	uss	$\sim$8.71cm
Charged xi	$\Xi^\pm$	dss	$\sim$4.91cm
Lambda	Λ^0	uds	$\sim$7.89cm
Charmed lambda	$\Lambda^\pm_c$	udc	$\sim$59.9 μm
Bottomed lambda	Λ^0_b	udb	$\sim$439 μm
Charged D meson	$D^\pm$	$\bar{c}d$	$\sim$312 μm
Neutral D meson	D^0	$\bar{c}u$	$\sim$123 μm
Strange D meson	$D^\pm_s$	$\bar{c}s$	$\sim$150 μm
Charged B meson	$B^\pm$	$b\bar{d}$	$\sim$491 μm
Neutral B meson	B^0	$b\bar{u}$	$\sim$455 μm
Strange B meson	$B^\pm_s$	$b\bar{s}$	$\sim$453 μm
Charmed B meson	B^0_c	$b\bar{c}$	$\sim$152 μm

performed with protons or atom nuclei.[16] However, many different types of hadrons can be directly seen in a detector; a list of hadrons with respective decay lengths is given in Table 2.5.

In addition to the valence quarks, other partons may exist in a hadron, namely the *sea* quarks and the gluons; their existence correspond to QCD fluctuations, and are permitted within the *Uncertainty Relation* $\Delta E \Delta t \geq \hbar/2$. The content of hadrons is described in terms of the aforementioned PDFs, and changes as a function of the energy scale at which it is considered. In Fig. 2.2, the contributions from valence

---

[16]One should still mention the pion-pion and proton-pion scatterings [44].

and sea quarks are separated; since only two up valence quarks and only one down valence quark exist, the following sum rules apply:

$$\int_0^1 f_{u/p}(x)\, dx = 2 \quad \text{and} \quad \int_0^1 f_{d/p}(x)\, dx = 1 \qquad (2.40)$$

In addition, the momentum sum rule must apply at any scale:

$$\int_0^1 dx \sum_{i\in\{q,g\}} x f_{i/p}(x, Q^2) = 1 \qquad (2.41)$$

2.3.2.2 Jets

Hadrons themselves are often found collimated in *jets*; this can be best illustrated by the two *event displays* in Fig. 2.3, where the final states of proton–proton collisions at CMS are shown; jets may be seen as the experimental pendants to the partons in the final state of a strong interaction. If jets can be defined both at parton level or at hadron level, there is no unique way of clustering hadrons into jets: clustering is partly arbitrary. To define a *jet clustering algorithm*, one takes several important properties into account:

– The boundaries of a jet are in principle undefined.
– An algorithm has to be IRC, i.e. insensitive toward the emission of a soft gluon or of a collinear gluon.
– Small-size jets may miss important contributions from the showering of a hard parton, but large-size jets will be contaminated by extra activity in the event.

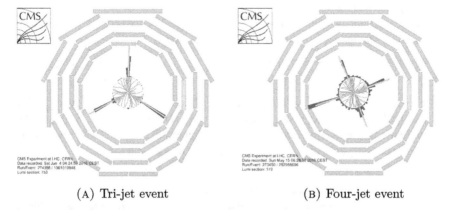

(A) Tri-jet event   (B) Four-jet event

Fig. 2.3 These two event displays show the final states of proton–proton collisions at CMS seen in the transverse plane [45]. In the innermost part, the green lines represent the reconstructed tracks corresponding to charged particles. The red and blue piles correspond to the energy deposits from all particles except muons and neutrinos; they are the signature of jets

Different algorithms have been developed: the cone and the recombination algorithms [46, 47]. The difficulty in defining jets is that they need to be convenient both in predictions and in measurements [48].

Cone. These algorithms are defined in the (y, ϕ) space with rigid boundaries, where y (ϕ) stands for the rapidity[17] (azimuthal angle). Cone algorithms present the advantage of being easy to implement, but suffer from the difficulty of dealing with overlapping jets. Today, the most cited cone algorithm is the SIScone (Seedless and Infrared Safe Cone) algorithm[18] [49]; it is represented in the top left of Fig. 2.4. It presents certain advantages, as being IRC. However, the SIScone algorithm still relies on the choice of a non-physical parameter f to deal with overlapping cones. In addition, in comparison with the recombination algorithms (defined in the next paragraph), the SIScone is very time- and resource-consuming.

Recombination algorithms. All the recombination algorithms have the advantage of being IRC safe, and may be seen a particular case of the following algorithm:

- Define the distances d_{ij} between any two particles i and j of transverse momentum $k_{\perp i,j}$:

$$d_{ij} = \min \left(k_{\perp i}^{2p}, k_{\perp j}^{2p} \right) \frac{\Delta y_{ij}^2 + \Delta \phi_{ij}^2}{R^2} \qquad (2.42)$$

  where R is the cone size radius parameter, and for p the exponent parameter.
- Define the distances d_{iB} between any particle i and the beam B:

$$d_{iB} = k_{\perp i}^{2p} \qquad (2.43)$$

- Then one proceeds iteratively:
  1. Find first the minimum of the entire set of distances d_{ij}, d_{iB}.
  2. If d_{ij} is smaller, than cluster i and j into a (proto) jet by summing their momenta; if d_{iB} is smaller, then label i as a jet.
  3. If all particles have not been assigned to a jet, redefine the entire set of distances with the new objects and return to 2; else continue.

These algorithms are obviously invariant under boosts along the z-axis, along which the incident protons scatter. One should mention the three most frequently used ones:

k_T The k_T algorithm [50] (bottom left) is obtained for $p = 1$. This algorithm assume that the particles inside of a jet should have similar momenta. Soft particles are first clustered, explaining the irregular shape; this shape is very sensitive to the surrounding activity (the Underlying Event and the pile-up).

---

[17] See Appendix 2.B for the description of the coordinates.

[18] One can also mention the Iterative Cone algorithm with Progressive Removal procedure (collinear unsafe), or the Iterative Cone algorithm with Split Merge procedure (infrared unsafe). These two algorithms suffer from the fact that they rely on the somewhat arbitrary choice of a seed, i.e. a particle that would define the direction of the cone. The seeds lead to several ill-defined behaviours; the SIScone algorithm does not need any seed.

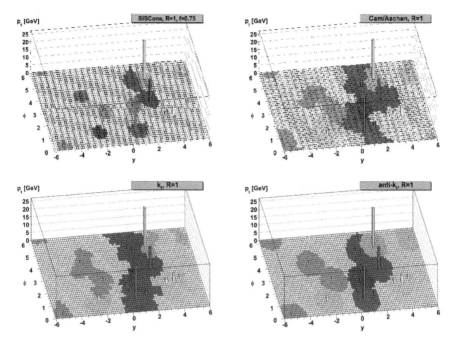

Fig. 2.4 Illustration of the four jet clustering algorithms with $R = 1$ for the same event from a simulation. Figure reproduced from authors [52]

Cambridge-Aachen The Cambridge-Aachen algorithm [51] (top right) is obtained for $p = 0$. The jet is also very sensitive to the surrounding activity but the substructure of the jet is totally conserved.

anti-k_T The anti-k_T algorithm [52] (bottom right) is obtained for $p = -1$. This algorithm is similar to the k_T algorithm but clusters first hard particles, which is related to the regular shape; this shape is robust against the surrounding activity.

As an illustration of the difference between the clustering algorithms, Fig. 2.4 shows the result of four different algorithms of jet clustering for the same event. At LHC Run-II, unless one is interested in the jet substructure, the standard algorithm used at CMS and ATLAS is the anti-k_T algorithm with cone size radius $R = 0.4$ or $R = 0.8$, in order to ease the comparison of results.

In this thesis, only one jet algorithm is considered: the standard anti-k_T algorithm shall be used with $R = 0.4$.

2.3.3 Overview of a Proton–Proton Scattering at LHC

The overview will follow the sketch in Fig. 2.5; the goal is to go through the different components:

Fig. 2.5 Sketch of a hadron-hadron collision as simulated by a Monte Carlo event generator. The red blob in the center represents the hard collision, surrounded by a tree-like structure representing Bremsstrahlung as simulated by parton showers. The purple blob indicates a secondary hard scattering event. Parton-to-hadron transitions are represented by light green blobs, dark green blobs indicate hadron decays, while yellow lines signal soft photon radiation. The incident partons, after extraction from the PDFs, are in blue. Finally, multi-parton interaction are shown in cyan. Figure made by Frank KRAUSS and reproduced with permission [53]

1. Matrix Element (ME),
2. Parton Shower (PS),

    - Initial-State Radiation (ISR),
    - Final-State Radiation (FSR);

3. Multi-Parton Interaction (MPI) and Beam–Beam Remnants (BBR);
4. hadronisation;
5. stable particles,

    - hadron decays,
    - soft photon radiation.

This description corresponds to the current understanding, as implemented in the General-Purpose MC Event Generators, whose implementation will be further discussed in Chap. 4.

2.3.3.1 Hard Process

The starting point of the overview of the proton–proton scattering is the factorisation (Eq. 2.37).

First, the hard process (represented with a red blob in the figure) can be described in the perturbative regime, as introduced in Sect. 2.2. As of today, predictions can be performed typically at LO or NLO, and in some cases at higher orders.

The hard process defines the *hard scale* Q^2 of the process. At LO, a natural choice corresponds to taking the virtuality of the internal propagator; at higher orders, there is no obvious way to proceed; in general, the choice of the hard scale is matter of debate.

The hard scale is then used as factorisation scale, i.e. $\mu_F^2 = Q^2$, at which the PDFs have to be considered. In order to reach the right scales, the PDFs of the incident hadron (three green lines with a green blob) can be evolved as a function of the scale. This is achieved with the DGLAP equations:

$$\frac{\mathrm{d}f_a(x, \mu_F^2)}{\mathrm{d}\ln \mu_F^2} = \sum_{b \in \{q, g\}} \int_x^1 \frac{\mathrm{d}z}{z} \frac{\alpha_S}{2\pi} P_{ba}(z) f_b\left(\frac{x}{z}, \mu_F^2\right) \tag{2.44}$$

The (collinear) splitting functions P_{ba}s describe the transition (after emitting one or several partons) of a parton b into a parton a carrying a momentum fraction z of the initial parton. Their exact expressions depend on the order of precision in which the evolution is performed.

At LO, they can be deduced by comparing the $2 \to 2$ and $2 \to 3$ MEs [22, 54] and, for massless quarks, correspond to the following expressions:

$$P_{qq}^{\mathrm{LO}}(z) = \frac{4}{3}\frac{1+z^2}{1-z} \tag{2.45}$$

$$P_{gq}^{\mathrm{LO}}(z) = \frac{3}{2}\left(z^2 + (1-z)^2\right) \tag{2.46}$$

$$P_{qg}^{\mathrm{LO}}(z) = \frac{4}{3}\frac{1+(1-z)^2}{z} \tag{2.47}$$

$$P_{gg}^{\mathrm{LO}}(z) = 3\left(\frac{z}{1-z} + \frac{1-z}{z} + z(1-z)\right) \tag{2.48}$$

In order to deal with the divergences in the evolution, a more rigorous expression
of the splitting functions should include some regularisation [53]; since it is not
useful for the present discussion, the regularisation is neglected for clarity. Moreover,
additional splitting functions also exist to include the photons and leptons in the
evolution; again for clarity, we restrict the discussion to QCD.

The interpretation of DGLAP equations (Eq. 2.44) is the following: when evolving
to a higher scale, the PDF must account for additional splittings, i.e. finer fluctuations
can be resolved. Furthermore the evolution can only be applied in the perturbative
regime; a PDF cannot be evolved lower than the hadronisation scale. In other words,
the DGLAP equations only describe the evolution of the perturbative component of
the PDF.

The interpretation in terms of fluctuations explains how non-u and non-d quarks
may appear. However, it may not always be correct to neglect the masses of
HF quarks, since these belong to scales of the perturbative regime. Today, different *flavour schemes* exist, according to whether they are treated as massive or as
massless:

1. In the *5-flavour scheme*, the b quark is treated as light and c quarks, i.e. it comes
   from PDFs; b quarks may be produced already at LO, but will be massless as any
   other quarks (except top).
2. In the *4-flavour scheme*, $b\bar{b}$ pairs can only be produced explicitly in the ME; in
   that case, calculations including higher-order terms are required, but on the other
   hand the mass of b's can be included.
3. Similarly, one can also define a *3-flavour scheme*.

Unlike all other quarks, the top quark is always considered as massive.

2.3.3.2   Parton Shower

When applying scrupulously the factorisation and using a fixed-order ME, the extra
ISR is not described explicitly, but only accounted for in the PDFs; however, radiations in the perturbative regime (namely the *shower*) should be *resolvable* in the
detector (red an blue lines). Moreover, additional FSR is also expected to take place,
which apply to legs originating either from the hard process of from ISR legs.

Figure 2.6a illustrates the PS in a $gg \rightarrow gg$ scattering, showing a realistic scenario
of the expected multiplicity, while the NLO or even NNLO calculations can only treat
up to a few legs. Therefore, in order to obtain a description of the high multiplicity
in the final state, one resorts to the evolution equations again [22, 53, 55, 56]. The
formulation of the DGLAP equations in Eq. 2.44 is *inclusive* in the sense that it only
allows to change the scale of the PDF without describing explicitly *when* a branching
occurs; for this, it is necessary to rewrite it in an *exclusive* form.

The evolution can be performed iteratively from the hard scale of the hard process
down to the hadronisation scale $Q_0 \sim \Lambda_{\text{QCD}}$. The treatment of ISR and FSR is similar,
despite some differences:

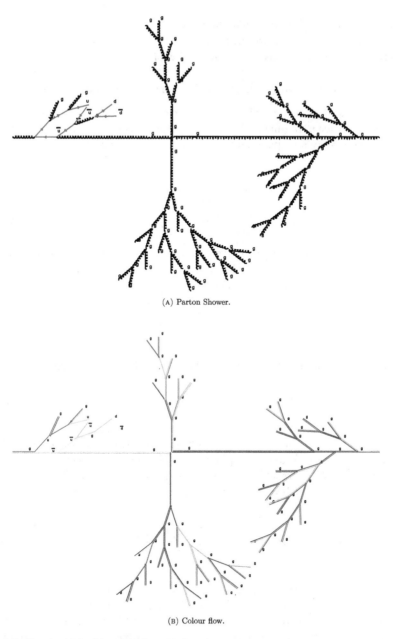

(A) Parton Shower.

(B) Colour flow.

Fig. 2.6 Sketch of PS with colour flow. The two incident (product) hard gluons are represented horizontally (vertically); the other lines corresponds to the result of showering. Above, the lines respect the representation in Feynman diagrams; below, the colour flow in the large-N limit is depicted for the same event. Figures made by Radek ŽLEBčÍK and reproduced with permission

- First, the ISR (FSR) is space-like (time-like).
- Then, the ISR is evolved *backwards* from the ME back to initial hadron (though only for efficiency reasons), while the FSR evolves without such constraint since the hadronic final state is *a priori* not known; in fact, in the context of FSR, the DGLAP evolution involves Fragmentation Functions (FFs) instead of PDFs.

Final-State Radiation. The FSR is the simpler and therefore discussed first. In order to determine when a splitting takes place, the DGLAP equations are reformulated in terms of probability for a parton a to undergo a splitting and become a parton b:

$$d\mathbb{P}_a(z, Q^2) = \frac{d f_a(z, Q^2)}{f_a(z, Q^2)} \tag{2.49}$$

$$= d \ln Q^2 \frac{\alpha_S}{2\pi} P_{ab}(z) \, dz \tag{2.50}$$

The probability of branching at a given scale regardless of the momentum fraction is described by the integral of Eq. 2.50 over all resolvable values of $z_{min} < z < z_{max}$ (which excludes de facto the divergences):

$$I_{ab}(Q^2) = \int_{z_{min}}^{z_{max}} dz \frac{\alpha_S}{2\pi} P_{ab}(z) \tag{2.51}$$

For an infinitesimal evolution $\delta \ln Q^2$, the probability for branching is given as follows:

$$\mathbb{P}_a(Q^2, Q^2 + \delta Q^2) = \sum_{b \in \{q, g\}} I_{ab}(Q^2) \delta \ln Q^2 \tag{2.52}$$

The probability for no branching is given by the complementary probability. Then, for an evolution from Q_1^2 to Q_2^2, the probability for no branching is given by a product of probabilities of not branching on infinitesimal intervals; eventually, it is given by an exponential and is called *Sudakov (form) factor*:

$$\Delta_a^{FSR}(Q_1^2, Q_2^2) \equiv \bar{\mathbb{P}}_a(Q_1^2, Q_2^2) = \exp\left(-\int_{Q_1^2}^{Q_2^2} d \ln Q^2 \sum_{b \in \{q, g\}} I_{ab}(Q^2)\right) \tag{2.53}$$

where $\bar{\mathbb{P}}_a$ stands for the probability for no branching. It can be interpreted in analogy with the decay of a particle, described with a Poisson process, $P(t) = \exp(-Nt)$ by identifying the (logarithm of the) scale $\ln Q^2$ with the time and the probability of branching with the decay rate N, with the only difference that in the case of the PS, the decay rate depends on the time. Moreover, a branching can take place on any outgoing leg of a diagram (unless this leg has already reached the hadronisation scale); the solution for the dependence in "time", or more exactly in the scale, is to proceed iteratively with a Sudakov factor associated to all outgoing legs:

1. Start from the scale of the hard process.
2. The Sudakov factors determine the scale of next branching (at a lower scale).
3. The branching of highest scale defines the new scale; one leg is added to the diagram.
4. If the scale has not reached the hadronisation scale yet, the procedure is iterated from Item 2; otherwise, stop the evolution.

The implementation is further described in Chap. 4. At each branching, the conservation laws are applied (four-momentum and quantum numbers in general).

Initial-State Radiation. The treatment of ISR is similar, with the difference that since the evolution is performed backwards, the PDF has to be taken into account explicitly in the probability of branching:

$$\frac{d f_b(x, Q^2)}{d f_b(x, Q^2)} = d \ln Q^2 \frac{\alpha_S}{2\pi} \int_x^{z_{max}} \frac{dz}{z} \frac{f_a(x/z, Q^2)}{f_b(x, Q^2)} P_{ab}(z) \qquad (2.54)$$

This results in a Sudakov factor weighted by the PDF, therefore also keeping a dependence in the momentum fraction carried by the parton:

$$\Delta_b^{ISR}(x, Q_1^2, Q_2^2) = \exp\left(-\int_{Q_1^2}^{Q_2^2} d \ln Q^2 \int_x^{z_{max}} \frac{dz}{z} \frac{f_a(x/z, Q^2)}{f_b(x, Q^2)}\right) \qquad (2.55)$$

Given this difference, the procedure is the same.

Strong ordering. The addition of a leg to the cross section formally corresponds to the following factorisation of the cross section:

$$d\sigma_{F+k+1} = d\sigma_{F+k} \Delta_a(Q_k^2, Q_{k+1}^2) \frac{\alpha_S}{2\pi} P_{ab}(z) \, dz \, d \ln Q^2 \qquad (2.56)$$

In principle, adding a diagram with one more leg gives rise to an interference term; it can be shown that this interference can be neglected in the context of *strong ordering* of the scale:

$$Q_{hadronisation}^2 \ll \cdots \ll Q_{F+k+1}^2 \ll Q_{F+k}^2 \ll \cdots \ll Q_{hard\,process}^2 \qquad (2.57)$$

Moreover, several possibilities exist for the scale, and may differ in the ISR compared to the FSR [56].

- The most natural choice consists in identifying the scale with the virtuality p^2 of the particles [57].
- An equivalent choice is to consider the *angular ordering*, since $dQ^2/Q^2 = d\theta/\theta$ [55].
- Finally, another possibility is to use transverse-momentum ordering [56].

In general, any scale such that $Q^2 = f(z)p_t^2$ with f a "reasonable" function can hold and p_t the transverse momentum [56].

Intrinsic k_T. In order to reflect the Fermi motion inside of the proton, an *intrinsic* k_T, or *primordial* k_T, is given to the initial partons, such that $k_T \sim \mathcal{O}(100\,\text{MeV})$. Conceptually, this is related to the fact that the DGLAP evolution only applies to the perturbative regime and treats the partons in a purely collinear way. In practice, a Gaussian is used to describe the primordial k_T. Then, the successive branchings make the partons acquire larger and larger transverse momenta.

Parameters. The PS comes with a few parameters, typically

- the strong coupling α_S,
- the hadronisation scale Q_0,
- the maximum momentum fraction $z_{\max}$,
- and the width σ of the Gaussian for the intrinsic k_T.

2.3.3.3 Multi-parton Interactions and Beam–Beam Remnants

In the strict application of the factorisation, the underlying colour interactions between the incident protons (purple blob and lines) is neglected. In this picture, the brute calculation of the cross section for $2 \rightarrow 2$ QCD processes is divergent:

$$\frac{\mathrm{d}\hat{\sigma}}{\mathrm{d}p_T} \propto \frac{\alpha_S(p_T^2)}{p_T^2} \tag{2.58}$$

Historically, the possibility of MPI was suggested by the excessive behaviour of the calculation for $p_T \sim 3-5\,\text{GeV}$, larger than the total cross section. In fact, additional hard interactions at similar scales may even take place; in that case, the formalism of Double-Parton Scattering (DPS) can be applied [58].

Smoothing factor. In order to cope with the divergent behaviour at low transverse momentum, the PS of a branching is then either interrupted below a certain value p_{T0} [55] or tamed with a smoothing factor in the cross section in Eq. 2.58 [56]:

$$F(p_T) = \left(\frac{\alpha_S(p_{T0}^2 + p_T^2)}{\alpha_S(p_T^2)} \frac{p_T^2}{p_{T0}^2 + p_T^2} \right)^2 \tag{2.59}$$

In both cases, the parameter p_{T0} has to be fitted from data.[19]

Number of interactions and interleaved Parton Shower. However, if Eq. 2.59 solves the divergence at low transverse momentum, it does not reproduce the multi-

---

[19]In fact, it is itself decomposed in further parameters with a power-low function of the centre-of-mass energy of the collisions:

$$p_{T0}\left(\sqrt{s}\right) = p_{T0}^{\mathrm{ref}} \left(\frac{\sqrt{s}}{\sqrt{s_0}} \right)^{\epsilon} \tag{2.60}$$

The three parameters p_{T0}^{ref}, $\sqrt{s_0}$ and ϵ can be determined with data samples from different experiments. This is not crucial for the current discussion but will be discussed in Part. II.

plicity in the final state of the event. Several MEs are used to describe the different hard scatterings, but share the same Underlying Event (UE):

- The average number of interactions can be estimated from the ratio of the hard and non-diffractive cross sections [57]:

$$\langle n \rangle = \frac{\sigma_{\text{hard}}}{\sigma_{\text{non-diff}}}$$

(2.61)

The number of interactions is follows a Poisson distribution.
- Since the different hard processes come from the same protons, the PS of the different hard processes must be *interleaved* [56]. The same procedure is applied, but all outgoing legs are taken into account simultaneously to perform the shower.

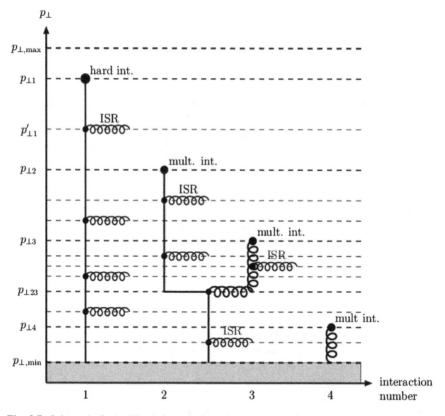

Fig. 2.7 Schematic figure illustrating one incoming hadron in an event with a hard interaction occurring at p_{T1} and three further interactions at successively lower p_T scales, each associated with (the potentiality of) initial-state radiation, and further with the possibility of two interacting partons (2 and 3 here) having a common ancestor in the parton showers. Full lines represent quarks and spirals gluons. The vertical p_T scale is chosen for clarity rather than realism; most of the activity is concentrated to small p_T values. Figure reproduced with permission of authors [56]

This is illustrated with p_T-ordering with four interactions in Fig. 2.7. Actually, two hard scatterings may come from a common ancestor, as it is the case for the interaction 2 and 3 on the figure; this case is called *joined interactions* [56].

Colour reconnection. Another aspect of the MPI is the fact that the different interactions must share the same colour flow, which has an impact on the hadronisation process (described in the next subsection). A procedure to reconnect the branches of the different interactions can be applied [57]. A basic model of the colour reconnection simply consists in reconnecting partons randomly; but this is insufficient for two reasons: first gluons seem to be ordered so as to minimise colour interaction, secondly non-trivial correlations can still take place as well. A more advanced model, as will be considered in the analysis in Part II, accounts for all this in an iterative procedure where low-p_T interactions, ordered in scales, are given a probability to reconnect with the interaction of highest-p_T, defined as follows:

$$p_{\text{reconnect}}(p_T) = \frac{(R + p_{T0})^2}{(R + p_{T0})^2 + p_T^2} \tag{2.62}$$

where R is the free parameter of the model, and p_{T0} is taken as in Eq. 2.60.

Beam–Beam Remnants. Eventually, the dynamics of the spectator partons in the beam remnants needs to be treated for several reasons:

– consistent treatment of the colour flow in the hadronisation,
– compensation for the intrinsic k_T given to the hard partons,
– conservation of the quantum numbers.

The PDFs are adapted according to the partons that participated to the MPI: for n interactions, there can be up to $n + 3$ partons in the beam remnant; if a valence quark interacts, it is taken into account in the remaining valence band; if a sea quark interacts, it must leave a partner in the beam. Being coloured, the spectator partons are consequently responsible of a part of the hadronic activity in the forward region.

2.3.3.4 Hadronisation

Once the PS has stopped at a scale, the hadronisation takes place. In Fig. 2.6b, the *colour flow* after showering is illustrated. Partons are connected to one another and the hadronisation will transform the (coloured) partons to (white) hadrons. Given the chain of colour-connected partons, even causally separated partons may be involved in the same hadronisation process.

Formally, the hadronisation is described by FFs, which are analog to the PDFs for the transition from the perturbative regime to the non-perturbative regime. For processes defined by hadrons in the final state, the factorisation (Eq. 2.37) can be rewritten with additional FFs.

Two models exist: the *string model* and the *cluster model*, illustrated in Fig. 2.9 with e^+e^- scattering. Both treat colour in the large-N colour limit, where gluons may be considered as carrying one colour charge and one anti-colour charge.

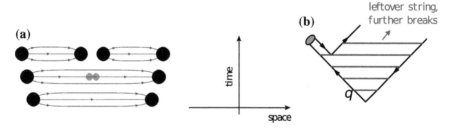

Fig. 2.8 **a** Illustration of string breaking by quark pair creation in the string field. **b** Illustration of the algorithmic choice to process the fragmentation from the outside-in, splitting off a single on-shell hadron in each step. Figures reproduced with permission of author [22]

Lund string fragmentation model. The interaction of any two connected partons is described by a *string* [57, 59]. Observations from Lattice QCD confirm that the colour field is concentrated in the form of a string [60]. In their centre-of-mass frame, the partons move apart from one another nearly at the speed of light; due to the property of confinement, the kinetic energy is transformed in potential energy; the potential energy is in turn converted into a $q\bar{q}$ pair of quarks, breaking the string by screening effect; this is illustrated in Fig. 2.8, where, on the left, the three stages from the string to the pair creations are shown, as initial quarks are getting further and further apart from one another, and on the right, the strings are represented and illustrate the screening effect at the production of a new pair. In the Lund fragmentation model, the FF are defined as follows:

$$f(z, m_T) = \frac{1}{z}(1-z)^a \exp\left(-b\frac{m_T^2}{z}\right) \tag{2.63}$$

where z is the longitudinal momentum fraction of the hadron from the quark, and $m_T^2 = m^2 + p_T^2$. The parameters a and b are not given by the theory and can depend on the flavour. One usually distinguishes different pairs of parameters (a, b) for light quarks (u, d, s), for charm quarks and for beauty quarks, but not for the top quark, since it is too heavy to participate to the hadronisation. Moreover, for HF quarks, an additional factor has been proposed [61]:

$$f_Q(z, m_T) = \frac{1}{z^{r_Q b m_Q^2}} \times f(z) \tag{2.64}$$

where Q is a generic notation for HF quark.

Cluster fragmentation model. Alternatively, the hadronisation is treated in smaller, white ensembles of partons with invariant mass in practice below 3 GeV, where the gluons are forced to split into quarks are forced to split into $q\bar{q}$ pairs (Fig. 2.9b) [55, 62, 63]. Similar dynamics as in the string fragmentation model is then adopted.

In the measurement presented in Part II, the final state is defined in terms of B hadrons rather than in terms of b quarks. Therefore we should in principle distinguish the notation of b jets and B jets; however, in order to stay consistent with the CMS convention [65], we shall keep the notation "b jets".

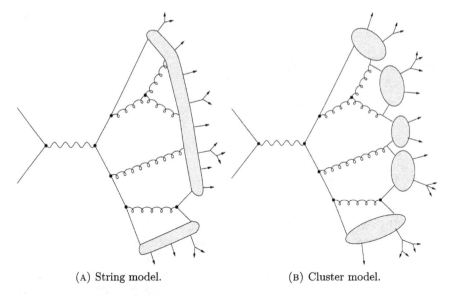

(A) String model.　　　　　　　　　(B) Cluster model.

Fig. 2.9 Sketch of the two main hadronisation models in *ep* scattering. The blobs correspond to the region where hadronisation processes take place. In the string model, far away separated partons may be taken into account in the same process; in the cluster model, smaller ensemble are first distinguished. Figures reproduced with permission of author [64]

2.3.3.5 Stable Particles

Hadrons differ significantly in terms of lifetime, which is important to treat in order to reflect a real event in a detector. In Table 2.5, the decay length of the most frequently produced hadrons are given, corresponding to the lifetime in the rest frame; in practice, these objects may be boosted, changing significantly the actual distance.

In parallel, the radiation of soft photons takes place.

2.3.4 More on Evolution

The discussion of the previous section was entirely conducted in the *collinear factorisation*, described with the DGLAP evolution. Albeit very successful for the description of many measurements, it is conceptually not completely satisfying, since it does not include a proper description of the transverse momentum (*cf.* intrinsic k_T in the PS).

In general, evolution equations can be seen as renormalisation equations for certain quantities like PDFs [54]. The DGLAP evolution equations are derived in the context of the collinear factorisation; other factorisation scheme exist, as will be described in this section (Fig. 2.10).

Fig. 2.10 Diagram of QCD evolution. The different schemes are represented: DGLAP describes evolution according to the scale Q^2; BFKL describes evolution according to the momentum fraction x; CCFM describes evolution as a function of both

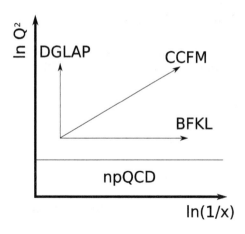

2.3.4.1 BFKL Evolution

The BFKL[20] [66–69] consists in an integro-differential equations in k_T and x for unintegrated gluon distribution G:

$$\frac{\mathrm{d}G(x, k_T^2)}{\mathrm{d} \ln 1/x} \sim \alpha_S \int \frac{\mathrm{d}q^2}{q^2} K\left(\frac{k_T^2}{q^2}\right) G(x, q^2) \tag{2.65}$$

where the unintegrated distribution can be related to the integrated gluon distribution by integrating over the transverse momentum k:

$$g(x, Q^2) = \int^{Q^2} \mathrm{d}^2 k\, G(x, k^2) \tag{2.66}$$

The condition for the application of this equation is a strong ordering in the successive momentum fractions, rather than in the scale. The situation can be seen in analogy to the DGLAP evolution where the rôles of x and Q^2 have swapped. At very low x, recombination effects from gluons are expected to become significant; this effect is called *saturation*. In case of saturation, BFKL does not apply anymore.

Experimental conditions to test BFKL equations require to go in the very forward region of the detector, because small x values means large rapidity separation:

$$\Delta y \sim \ln \frac{1}{x} \tag{2.67}$$

Experimental evidence was found with the so-called *Mueller-Navelet jets* [70]. In general, large rapidity separations will not be discussed in this thesis.

---

[20]Named after the four physicists BALITSKY, FADIN, KURAEV and LIPATOV.

2.3.4.2 CCFM Evolution

The CCFM[21] [71–74] evolution equation is also interesting since it includes a dependence in transverse momentum of PDFs:

$$f(x, Q^2) = \int dk_T \mathcal{A}(x, k_t, Q^2) \tag{2.68}$$

where f ($\mathcal{A}$) is the *collinear PDF* (TMD, or Transverse-Momentum-Dependent PDF).

In this approach, Eq. 2.37 can be explicitly written as follows:

$$\sigma_{h_1 h_2 \to X} = \sum_{a \in \{q,g\}} \int dx_a \, dk_{Ta} \mathcal{A}_{a \in h_1}(x_a, k_{Ta}, \mu_F^2) \times$$
$$\times \sum_{b \in \{q,g\}} \int dx_b \, dk_{Tb} \mathcal{A}_{b \in h_2}(x_b, k_{Tb}, \mu_F^2) \, d\sigma_{ab} \tag{2.69}$$

The angular ordering is applied. Nonetheless, only gluons are included in this evolution; therefore it is not suited for high transverse momentum regions, where quarks have a significant contribution, such as in this thesis.

Conclusion

In this chapter, the basic elements of theory to understand proton–proton collisions at LHC have been introduced. The complexity of performing predictions has been explained, and the difficulties inherent to hadrons, related to the property of confinement, has been presented. This portrait of HEP has been intended to be general.

Apart of the next chapter, dedicated to the description of the experimental set-up, we shall focus more on topics related to b physics, from generalities to the measurement of the production of B jets at hadron colliders.

In appendix of this chapter are a historical introduction to HEP (with a more theoretical point of view than in Chap. 1) and a description of some typical coordinates and variables used in the context of general-purpose hadron colliders.

---

[21] Named after the four physicists CATANI, CIAFALONI, FIORANI and MARCHESINI.

2.A Historical Perspective

Modern physics started with Galileo GALILEI, at the beginning of the 17th century, who first formulated the *Principle of Relativity*, stating that the laws of mechanics should be the same in all (inertial) frames [75].

The formulation of the three *laws of motion*[22] came up at the end of the 17th century with Isaac NEWTON [1]. These laws assume that the evolution of a particle follows a trajectory.

In parallel of his work on mechanics, NEWTON also wrote a treatise on *optics* [76], where he assumed the light of *corpuscular* nature.

At the beginning of the 18th century, Thomas YOUNG performed the *double-slit experiment* with lights, highlighting the wave nature of light, in contradiction with the idea of light made of classical corpuscles.

During the 19th century, James Clerk MAXWELL synthesised the *laws of electromagnetism* in four equations[23] [77–79]. These four equations had two consequences: first, solutions to the Maxwell equations led to a description of light in terms of electromagnetic waves, confirming YOUNG's interpretation; secondly, these equations did not respect the Principle of Relativity as expressed by GALILEI.

At the end of the 19th century, Hendrik LORENTZ and Albert EINSTEIN found a new formulation of the Principle of Relativity, where the speed of light c appears as a fundamental constant of physics, valid in all frames [80–82]; today, the "relativistic" vocable refers to this version of the Principle of Relativity. In addition, MAXWELL noticed that the electric and magnetic field were invariant under a *Gauge transformation*.[24]

---

[22]Personal translation from Latin:

Principle of Inertia: ≪ Every body will to stay in its state of rest or of uniformly straight motion, unless it is compelled to change its state by a force that applies on it.≫

Fundamental principle of dynamics: ≪ The change of the motion is proportional to the force that is applied on it, and is applied in straight line with respect to it. (...) ≫

Principe of reciprocal actions: ≪ The reaction is always equal and opposite to the action (...) ≫

[23]Given the electric charge density ρ and is the electric current density $\mathbf{j}$, the electric and magnetic fields $\mathbf{E}$ and $\mathbf{B}$ are described by the four following equations:

Gauss $\nabla \cdot \mathbf{E} = \rho/\epsilon_0$
Maxwell $\nabla \cdot \mathbf{B} = 0$
Faraday $\nabla \times \mathbf{E} = -\partial \mathbf{B}/\partial t$
Ampère $\nabla \times \mathbf{B} = \mu_0 \mathbf{j} + \epsilon_0 \mu_0 \partial \mathbf{E}/\partial t$

[24]The electric and magnetic fields can be deduced from a scalar and a vectorial potentials ϕ and $\mathbf{A}$:

$$\mathbf{E} = -\nabla\phi - \frac{\partial \mathbf{A}}{\partial t}, \quad \mathbf{B} = \nabla \times \mathbf{A} \qquad (2.70)$$

These potentials are not unique; certain transformations, called *Gauge transformations*, lead to the same evolution:

$$\phi \longrightarrow \phi + \frac{\partial \Lambda}{\partial t}, \quad \mathbf{A} \longrightarrow \mathbf{A} + \nabla\Lambda \qquad (2.71)$$

At the beginning of the 20th century, in parallel of the development of relativity, Max PLANCK, in his attempts to describe the *radiation of the black body*, first used discrete levels of energy [83–85]. Although he himself did not believe at first in it, it turned out that *quantisation* could explain several phenomena: for instance, the *photoelectric effect* [86] or the *spectral lines* with the atomic *Bohr model* [87, 88]. (The repetition of YOUNG's experiment with electrons or larger molecules, shown in Fig. 2.11, would come only in the second half of the 20th century [89, 90].) The general description of (non-relativistic) quantic states was obtained by Erwin SCHRÖDINGER[25] [91–95]; while NEWTON's laws of motion describe the evolution of a particle with a trajectory, the Schrödinger equation describes the evolution of a particle with a *wave function*, which is then understood as a probability amplitude. Experimentally, DAVISSON and GERMER highlighted the wave-like character of the motion of electrons [96], similarly to the light.

In the context of quantum mechanics, a new quantity, showing the properties of an intrinsic angular momentum, was discovered: the *spin*. The *Stern-Gerlach experiment* [98–100], illustrated in Fig. 2.12, showed that its value is non-integer; it would turn out that it can take any half-integer for values. In a phenomenological approach, Wolfgang PAULI formulated an equation[26] with a 2-component wave function, called a *spinor*.

The direct generalisation of the Schrödinger equation to the relativistic regime is called the Klein-Gordon equation.[27] Unlike the Schrödinger equation, the interpretation of the wave function as a probability amplitude was not so clear anymore, because the conserved quantity—which corresponds to the probability in the case of the Schrödinger equation—was no more positive definite. Moreover the Klein-Gordon equation is only able to describe the dynamics of spinless particles, which does not include particles like the electron; experimental possibilities were limited at the time of its derivation.

---

for any function Λ.

[25] Schrödinger equation:

$$\left(\frac{\hat{\mathbf{p}}}{2m} + \hat{\mathbf{U}}(t) \right) |\psi\rangle = E |\psi\rangle \tag{2.72}$$

[26] Pauli equation:

$$\left[\frac{1}{2m} \left(\boldsymbol{\sigma} \cdot (\mathbf{p} - q\mathbf{A}) \right)^2 + q\phi \right] |\psi\rangle = i \frac{\partial}{\partial t} |\psi\rangle \tag{2.73}$$

for a half-spin particle in an electric (magnetic) field ϕ (**A**), with Pauli matrices:

$$\sigma_1 = \begin{bmatrix} 0 & 1 \\ 1 & 0 \end{bmatrix}, \quad \sigma_2 = \begin{bmatrix} 0 & -i \\ i & 0 \end{bmatrix} \quad \sigma_3 = \begin{bmatrix} 1 & 0 \\ 0 & -1 \end{bmatrix} \tag{2.74}$$

[27] While the Schrödinger equation is built up on the dispersion relation $E = p^2/2m$, the Klein-Gordon equation is built up on its relativistic analog $E^2 = m^2 + p^2$: $\Box\phi + m^2\phi = 0$.

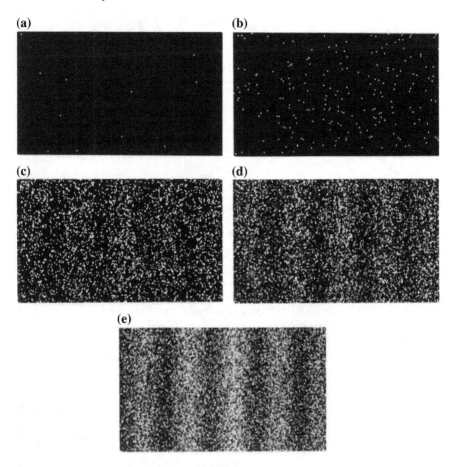

Fig. 2.11 Results of a double-slit-experiment performed by Dr. TONOMURA showing the build-up of an interference pattern of single electrons. Numbers of electrons are 11 (**a**), 200 (**b**), 6000 (**c**), 40000 (**d**), 140000 (**e**) [97]

Paul DIRAC was the first in 1928 to derive an equation[28] [102, 103] in the relativistic regime to describe (free) particles with half spin, e.g. electrons. Expecting 2-components wave functions like in the Pauli equation, DIRAC built up his equation in such a way that each component would respect the Klein-Gordon equation. However, it turned out that in order to fulfill all requirements, the wave function needed

---

[28] Given a four-component wave function ψ:

$$\left(i\gamma^{\mu}\partial_{\mu} - m\right)\psi = 0 \tag{2.75}$$

where

$$\gamma^{0} = \begin{bmatrix} 1_2 & 0 \\ 0 & 1_2 \end{bmatrix}, \quad \gamma^{i} = \begin{bmatrix} 0 & -\sigma_i \\ \sigma_i & 0 \end{bmatrix} \tag{2.76}$$

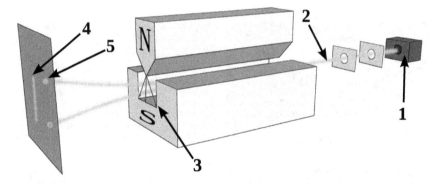

Fig. 2.12 Stern-Gerlach experiment: silver atoms travel through an inhomogeneous magnetic field and are deflected up or down depending on their spin. *1* furnace. *2* beam of silver atoms. *3* inhomogeneous magnetic field. *4* expected result. *5* what was actually observed [101]

not two but four components; in addition, some solutions seemed to correspond to negative energies. By introducing an *ad hoc* term to describe interactions with an electromagnetic field,[29] one could find in the non-relativistic limit the Pauli equation for two of the four components; the two additional components suggested the existence of *antiparticles*, soon confirmed by Carl David ANDERSON with positrons [104].

Eventually, the interpretation of the wave function was abandoned—some solutions had negative energies, and the causality was not respected—for that of *quantum field*; a quantic and relativistic theory must fundamentally treat of multi-particle systems. Instead of describing the quantic particle itself and its probability amplitude of being in a given state, the quantum field describes creation and annihilation of particles and antiparticles. Particles are only excitation states of the quantum field, which fills up the whole space.

Soon after, the Proca equation,[30] the EOM for spin-1 particles, came up, as well as the *spin-statistics theorem* [105] establishing the relations between the spin and the statistics of the particles; one could therefore distinguish two types:

fermions   half-integer-spin particles (obeying the *Pauli Exclusion Principle*);
bosons   integer-spin particles (which can be superposed arbitrarily).

In parallel, after the discovery of the neutron by James CHADWICK [106], Hideki YUKAWA developed a theory of nuclear interaction.

---

[29] Given a four-component wave function ψ:

$$\left(\gamma^\mu \left(i\partial_\mu - A_\mu\right) - m\right)\psi = 0 \tag{2.77}$$

[30]

$$\partial_\mu F^{\mu\nu} + m^2 A^\nu = 0 \tag{2.78}$$

The next decades would show the discovery of many new particles and phenomena. Before closing this section, two of them at least should still be mentioned: first, the experimental discovery of partons, predicted by GELL- MANN and FEYNMAN at Stanford National Accelerator Laboratory (SLAC) in 1968 (see Sect. 1.2.4) [107], and then the discovery of the colour charge with the Δ^{++}. This marked the day of birth of the QCD, which we are treating in this thesis.

The SM was developed in the following decades with growing successes. Today, all particles predicted by the SM have been discovered, all parameters have been measured. Still, many theoretical questions stay open, like the group structure of SM and some apparent symmetries.

2.B Variables and Coordinates

We recapitulate here some of the standard variables and coordinates used in HEP.

2.B.1 Coordinate System

We adopt here the conventional coordinate system used at CMS, as shown in Fig. 2.13a:

– The x axis points toward the centre of ring of the collider.

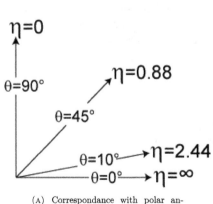

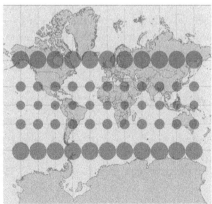

(A) Correspondance with polar angle [108].

(B) Mercator projection of the planet Earth [109]: the red disks correspond to disks of same area on the globe, illustrating the effect of the conformal transformation in Eq. 2.79.

Fig. 2.13 Conventional coordinates at CMS. z is the axis of the beam, $\eta = -\ln\tan\frac{\theta}{2}$, ϕ the azimuthal angle

– The y axis points to the sky.
– The z axis is defined along the colliding beams with the right hand rule.

Given the symmetry of the detectors and of the topology of the collisions, cylindrical coordinates are often used.

However, instead of using a polar angle θ, one uses *pseudorapidity*:

$$\eta = -\ln \tan \frac{\theta}{2} \tag{2.79}$$

The pseudorapidity η is convenient since the difference of pseudorapidities $\Delta\eta$ are *almost* Lorentz-invariant quantity. The pseudorapidity also corresponds to the Mercator projection used for maps of the Earth (see Fig. 2.13b).

The *rapidity* y is defined as follows:

$$y = \frac{1}{2} \ln \frac{E + p_z}{E - p_z} \tag{2.80}$$

In cylindrical coordinates, the difference of rapidities Δy is Lorentz invariant along the z axis, and can be related to the polar angle for small masses, i.e. $m \ll |\mathbf{p}|$:

$$
\begin{aligned}
y &\approx \frac{1}{2} \ln \frac{|\mathbf{p}| + p_z}{|\mathbf{p}| - p_z} \\
&= \frac{1}{2} \ln \frac{1 + \cos\theta}{1 - \cos\theta} \\
&= -\ln \tan \frac{\theta}{2} \equiv \eta
\end{aligned}
$$

The transformation from the polar angle to the pseudorapidity is a conformal transformation, which means that angles are conserved and that a circle is transformed in a circle; it is therefore well suited for the definition of jet cone.

2.B.2 Mandelstam Variables

In the case of $2 \rightarrow 2$ processes, it is convenient to define the *Mandelstam variables*. The Mandelstam variables correspond to three Lorentz-invariant quantities with interesting properties.

2.B.2.1 Definition

The Mandelstam variables are defined as follows:

$$s = (p_1 + p_2)^2 = (p_3 + p_4)^2 \tag{2.81}$$

$$t = (p_1 - p_3)^2 = (p_4 - p_2)^2 \tag{2.82}$$
$$u = (p_1 - p_4)^2 = (p_2 - p_3)^2 \tag{2.83}$$

For instance, s coincides with the squared centre-of-mass energy of a collision, and is an important parameter for colliders; t coincides with the *momentum transfer* of a collision.

2.B.3   *Channels*

The Mandelstam variables are also used to designate the three channels at tree level:

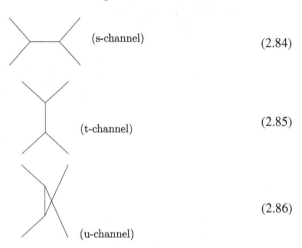

(s-channel)   (2.84)

(t-channel)   (2.85)

(2.86)

(u-channel)

The permutation of two Mandelstam variables corresponds to the permutation of legs in a diagram; the MEs of different processes may be easily related thanks to this property.

Parton convention. If the scattering involves hadrons, the Mandelstam variables at the parton-level (hadron-level) are used with (without) hat. In the collinear factori-sation, the s variables at hadron and parton levels can be related with the momentum fractions; for DIS and pp scatterings respectively:

$$\hat{s} = x_1 s \qquad\qquad \text{(DIS)} \qquad\qquad (2.87)$$
$$\hat{s} = x_1 x_2 s \qquad\qquad (pp) \qquad\qquad (2.88)$$

2.B.4   *Properties*

General. Let m_i be the masses of the interacting particles:

$$s + t + u = \sum_{i=1}^{4} m_i^2 \tag{2.89}$$

Centre-of-mass frame. The t and u variables can be related to the polar angle in the centre-of-mass (c.m.s.) frame.

$$t = \frac{-s}{2} (1 - \cos\theta) \tag{2.90}$$

$$u = \frac{-s}{2} (1 + \cos\theta) \tag{2.91}$$

Massless approximation. If the mass can be neglected, the form of the Mandelstam variables is simplified to the following:

$$s \approx 2p_1 \cdot p_2 \tag{2.92}$$

$$t \approx -2p_1 \cdot p_3 \tag{2.93}$$

$$u \approx -2p_1 \cdot p_4 \tag{2.94}$$

References

1. Newton I (1687) Philosophiae naturalis principia mathematica. Wikimedia (ed). https://la.wikisource.org/wiki/Philosophiae_Naturalis_Principia_Mathematica
2. Dissertori G, Knowles IG, Schmelling M (2003) Quantum chromodynamics: high energy experiments and theory, vol 115. Oxford University Press, Oxford
3. Halzen F, Martin AD (1984) Quarks and leptons: an introductory course in modern particle physics
4. Peskin ME, Schroeder DV (1995) An introduction to quantum field theory. Addison-Wesley, Reading. http://www.slac.stanford.edu/~mpeskin/QFT.html
5. Hambye T (2013) Introduction au modèle standard. Course of 1st master year, ULB. No written version
6. Hambye T, Tytgat MHG (2012) An introduction to elementary particles. Theoretical aspects of the course of 3rd bachelor year, ULB
7. Henneaux M, Groupes et représentations. http://www.solvayinstitutes.be/pdf/Marc/SO3SO3_1_18112014.pdf.b Lectures at ULB
8. Henneaux M, Représentation des groupes et application à la physique. http://www.solvayinstitutes.be/pdf/Marc/Groupes22062015.pdf. c Lectures at ULB
9. Felsager B (1981) Geometry, particles and fields. Graduate texts in contemporary physics. University Press, Odense. https://doi.org/10.1007/978-1-4612-0631-6
10. Yang CN (1954) Mills RL (1954) Conservation of isotopic spin and isotopic gauge invariance. Phys Rev 96:191–195. https://doi.org/10.1103/PhysRev.96.191
11. Howard G (1999) Lie algebras in particle physics. Front Phys 54:1–320
12. Gerardus't H (1971) Renormalization of massless Yang-Mills fields. Nucl Phys B 33(1):173–199
13. Barnich G, Localité et symétries en théorie classique des champs. http://homepages.ulb.ac.be/

14. Kobayashi M, Maskawa T (1973) CP-violation in the renormalizable theory of weak interaction. Prog Theor Phys 49(2):652–657. https://doi.org/10.1143/PTP.49.652
15. Tanabashi M et al (2018) Review of particle physics. Phys Rev D 98(3):030001
16. Ellis RK, Stirling WJ, Webber BR (2003) QCD and collider physics. Cambridge University Press, Cambridge
17. Serguei C et al (2012) Observation of a new boson at a mass of 125 GeV with the CMS experiment at the LHC. Phys Lett B 716(1):30–61
18. Georges A et al (2012) Observation of a new particle in the search for the Standard Model Higgs boson with the ATLAS detector at the LHC. Phys Lett B 716(1):1–29
19. Englert F, Brout R (1964) Broken symmetry and the mass of gauge vector mesons. Phys Rev Lett 13:321–323. https://doi.org/10.1103/PhysRevLett.13.321
20. Higgs PW (1964) Broken symmetries and the masses of gauge bosons. Phys Rev Lett 13:508–509. https://doi.org/10.1103/PhysRevLett.13.508
21. Favart L (2013) Physique auprès des collisionneurs. Course of 1st master year, ULB
22. Skands P, Introduction to QCD. In: Searching for new physics at small and large scales, pp 341–420. https://doi.org/10.1142/9789814525220_0008
23. Britzger D et al (2017) Determination of the strong coupling constant from inclusive jet cross section data from multiple experiments. arXiv:1712.00480 [hep-ph]
24. Feynman RP (1949) Space-time approach to quantum electrodynamics. Phys Rev 76(6):769
25. Bohr H, Nielsen HB (1977) Hadron production from a boiling quark soup: a thermodynamical quark model predicting particle ratios in hadronic collisions. Nucl Phys B 128(2):275–293. ISSN:0550-3213. https://doi.org/10.1016/0550-3213(77)90032-3
26. Yoichiro N (1976) The confinement of quarks. Sci Am 235(5):48–63
27. Wilson KG (1974) Confinement of quarks. Phys Rev D 10:2445–2459. https://doi.org/10.1103/PhysRevD.10.2445
28. Gupta R (1997) Introduction to lattice QCD: course. In: Probing the standard model of particle interactions. Proceedings, summer school in theoretical physics, NATO Advanced Study Institute, 68th session, Les Houches, France, July 28–September 5, 1997. Pt. 1, 2, pp 83–219. arXiv:hep-lat/9807028 [hep-lat]
29. Rothe HJ (2005) Lattice gauge theories: an introduction. World Scientific, Singapore
30. Alwall J et al (2014) The automated computation of tree-level and next-to-leading order differential cross sections, and their matching to parton shower simulations. JHEP 07:079. https://doi.org/10.1007/JHEP07(2014)079, arXiv:1405.0301 [hep-ph]
31. Khachatryan V et al (2015) Pseudorapidity distribution of charged hadrons in proton - proton collisions at s=13 TeV. Phys Lett B 751(Supplement C):143–163. ISSN:0370-2693. https://doi.org/10.1016/j.physletb.2015.10.004, http://www.sciencedirect.com/science/article/pii/S0370269315007558
32. CMS Collaboration (2016) Measurement of pseudorapidity distributions of charged particles in proton–proton collisions at $\sqrt{s}$ = 13 TeV by the CMS experiment
33. Irving AC, Worden RP, Regge phenomenology. Phys Rep 34(3):117–231. ISSN:0370-1573. https://doi.org/10.1016/0370-1573(77)90010-2
34. Collins PDB (1977) An introduction to regge theory and high energy physics. Cambridge University Press, Cambridge
35. Collins JC, Soper DE, Sterman GF (1989) Factorization of hard processes in QCD. Adv Ser Direct High Energy Phys 5:1–91. https://doi.org/10.1142/9789814503266_0001, arXiv:hep-ph/0409313 [hep-ph]
36. Alekhin S, Blumlein J, Moch S (2014) The ABM parton distributions tuned to LHC data. Phys Rev D 89(5):054028. https://doi.org/10.1103/PhysRevD.89.054028, arXiv:1310.3059 [hep-ph]
37. Pumplin J et al (2002) New generation of parton distributions with uncertainties from global QCD analysis. JHEP 07:012. https://doi.org/10.1088/1126-6708/2002/07/012, arXiv:hep-ph/0201195 [hep-ph]
38. Cooper-Sarkar AM (2014) HERAPDF1.5LO PDF set with experimental uncertainties. PoS DIS2014:032

39. Harland-Lang LA et al (2015) Parton distributions in the LHC era: MMHT 2014 PDFs. Eur Phys J C75(5):204. https://doi.org/10.1140/epjc/s10052-015-3397-6, arXiv:1412.3989 [hep-ph]
40. Martin AD et al (2009) Parton distributions for the LHC. Eur Phys J C63:189–285. https://doi.org/10.1140/epjc/s10052-009-1072-5, arXiv:0901.0002 [hep-ph]
41. Ball RD et al (2013) Parton distributions with LHC data. Nucl Phys B867:244–289. https://doi.org/10.1016/j.nuclphysb.2012.10.003, arXiv:1207.1303 [hep-ph]
42. Ball RD et al (2015) Parton distributions for the LHC run II. JHEP 04:040. https://doi.org/10.1007/JHEP04(2015)040, arXiv:1410.8849 [hep-ph]
43. Abramowicz H et al (2015) Combination of measurements of inclusive deep inelastic $e^{\pm}p$ scattering cross sections and QCD analysis of HERA data. Eur Phys J C75(12):580. https://link.springer.com/article/10.1140%2Fepjc%2Fs10052-015-3710-4, arXiv:1506.06042 [hep-ex]
44. Madigozhin D (2007) Pion scattering lengths from NA48. Nucl Phys B - Proc Suppl 164(QCD 05):85–88. ISSN:0920-5632. https://doi.org/10.1016/j.nuclphysbps.2006.11.069, http://www.sciencedirect.com/science/article/pii/S0920563206009212
45. Sirunyan AM et al (2017) Azimuthal correlations for inclusive 2-jet, 3-jet, and 4-jet events in pp collisions at $\sqrt{s}$ = 13 TeV. arXiv:1712.05471 [hep-ex]
46. Atkin R (2015) Review of jet reconstruction algorithms. J Phys: Conf Ser 645(1):012008. http://stacks.iop.org/1742-6596/645/i=1/a=012008
47. Salam GP (2010) Towards jetography. Eur Phys J C67:637–686. https://doi.org/10.1140/epjc/s10052-010-1314-6, arXiv:0906.1833 [hep-ph]
48. Huth JE et al (1990) Toward a standardization of jet definitions. In: 1990 DPF summer study on high-energy physics: research directions for the decade (Snowmass 90) Snowmass, Colorado, June 25–July 13, 1990, pp 0134–136. http://lss.fnal.gov/cgi-bin/find_paper.pl?conf-90-249
49. Salam GP, Soyez G (2007) A practical seedless infrared-safe cone jet algorithm. JHEP 05:086. https://doi.org/10.1088/1126-6708/2007/05/086, arXiv:0704.0292 [hep-ph]
50. Stefano C et al (1993) Longitudinally-invariant k?-clustering algorithms for hadron-hadron collisions. Nucl Phys B 406(1–2):187–224
51. Dokshitzer YL et al (1997) Better jet clustering algorithms. J High Energy Phys 08:001
52. Cacciari M, Salam GP, Soyez G (2008) The Anti-k(t) jet clustering algorithm. JHEP 04:063. https://doi.org/10.1088/1126-6708/2008/04/063, arXiv:0802.1189 [hep-ph]
53. Höche S (2015) Introduction to parton-shower event generators. In: Proceedings, theoretical advanced study institute in elementary particle physics: journeys through the precision frontier: amplitudes for colliders (TASI 2014): Boulder, Colorado, June 2–27, 2014, pp 235–295. https://doi.org/10.1142/9789814678766_0005, arXiv:1411.4085 [hep-ph]
54. Jung H (2015) Lectures on QCD and Monte Carlo. http://desy.de/~jung/qcd_and_mc_2015/lecture-writeup.pdf.2015
55. Bahr M et al (2008) Herwig++ physics and manual. Eur Phys J C 58:639. https://doi.org/10.1140/epjc/s10052-008-0798-9, arXiv:0803.0883 [hep-ph]
56. Sjostrand T, Skands PZ (2005) Transverse-momentum-ordered showers and interleaved multiple interactions. Eur Phys J C39:129–154. https://doi.org/10.1140/epjc/s2004-02084-y, arXiv:hep-ph/0408302 [hep-ph]
57. Sjöstrand T, Mrenna S, Skands P (2006) PYTHIA 6.4 physics and manual. JHEP 05:026. https://doi.org/10.1088/1126-6708/2006/05/026, arXiv:hep-ph/0603175 [hep-ph]
58. Gunnellini P (2014) Study of double parton scattering using four-jet scenarios in proton-proton collisions at $\sqrt{s}$ = 7 TeV with the CMS experiment at the large Hadron Collider. PhD thesis. U. Hamburg, Dept Phys. https://doi.org/10.1007/978-3-319-22213-4,10.3204/DESY-THESIS-2015-010, http://pubdb.desy.de/search?of=hd&p=id:DESY-THESIS-2015-010
59. Andersson B et al (1983) Parton fragmentation and string dynamics. Phys Rep 97(2):31–145. ISSN:0370-1573. https://doi.org/10.1016/0370-1573(83)90080-7
60. Bali GS, Schilling K (1992) Static quark-antiquark potential: scaling behavior and finite-size effects in SU (3) lattice gauge theory. Phys Rev D 46(6):2636

61. Bowler MG (1981) e+ e- production of heavy quarks in the string model. Z Phys C 11:169. https://doi.org/10.1007/BF01574001
62. Amati D, Veneziano G (1979) Preconfinement as a property of perturbative QCD. Phys Lett B 83(1):87–92
63. Webber BR (1984) A QCD model for jet fragmentation including soft gluon interference. Nucl Phys B 238:492–528. https://doi.org/10.1016/0550-3213(84)90333-X
64. Webber B (2011) Parton shower Monte Carlo event generators. Scholarpedia 6(12). Revision #128236, p 10662. https://doi.org/10.4249/scholarpedia.10662
65. Sirunyan AM et al (2017) Identification of heavy-flavour jets with the CMS detector in pp collisions at 13 TeV. arXiv:1712.07158 [physics.ins-det]
66. Fadin VS, Kuraev EA, Lipatov LN (1975) On the pomeranchuk singularity in asymptotically free theories. Phys Lett 60B:50–52. https://doi.org/10.1016/0370-2693(75)90524-9
67. Kuraev EA, Lipatov LN, Fadin VS (1976) Multi-reggeon processes in the Yang-Mills theory. Sov Phys JETP 44:443–450. [Zh Eksp Teor Fiz 71, 840 (1976)]
68. Kuraev EA, Lipatov LN, Fadin VS (1977) The pomeranchuk singularity in nonabelian gauge theories. Sov Phys JETP 45:199–204. [Zh Eksp Teor Fiz 72, 377 (1977)]
69. Balitsky II, Lipatov LN (1978) The pomeranchuk singularity in quantum chromodynamics. Sov J Nucl Phys 28:822–829. [Yad Fiz 28, 1597 (1978)]
70. Mueller AH, Navelet H (1987) An inclusive minijet cross section and the bare pomeron in QCD. Nucl Phys B 282:727–744
71. Marcello C (1988) Coherence effects in initial jets at small Q2/s. Nucl Phys B 296(1):49–74
72. Catani S, Fiorani F, Marchesini G (1990) QCD coherence in initial state radiation. Phys Lett B 234(3):339–345
73. Catani S, Fiorani F, Marchesini G (1990) Small-x behaviour of initial state radiation in perturbative QCD. Nucl Phys B 336(1):18–85
74. Giuseppe M (1995) QCD coherence in the structure function and associated distributions at small x ★. Nucl Phys B 445(1):49–78
75. Galilei G (1632) Dialogo sopra i due massimi sistemi del mondo tolemaico e copernicano. Wikimedia (ed) https://it.wikisource.org/wiki/Dialogo_sopra_i_due_massimi_sistemi_del_mondo_tolemaico_e_copernicano
76. Opticks NI (ed) by Wikimedia https://en.wikisource.org/wiki/Opticks_(4th_Ed).1704
77. Maxwell JC (1864) On Faraday's lines of force. Trans Camb Philos Soc 10:27
78. Maxwell JC (1864) A dynamical theory of the electromagnetic field. The society
79. Maxwell JC (1881) A treatise on electricity and magnetism, vol 1. Clarendon Press
80. Lorentz HA (1898) Simplified theory of electrical and optical phenomena in moving systems. Koninklijke Nederlandse Akademie van Wetenschappen Proc Ser B Phys Sci 1:427–442
81. Lorentz HA (1904) Electromagnetic phenomena in a system moving with any velocity smaller than that of light. KNAW Proc 6:1903–1904
82. Einstein A (1905) Zur Elektrodynamik bewegter Körper. Annalen der Physik 322(10):891–921. ISSN:1521-3889. https://doi.org/10.1002/andp.19053221004
83. Planck MKEL (1900) Zur theorie des gesetzes der energieverteilung im normalspectrum. Verhandl Dtsc Phys Ges 2:237
84. Planck M (1910) Zur Theorie der Wörmestrahlung. Annalen der Physik 336(4):758–768
85. Planck M (1921) Vorlesungen über die Theorie der Wärmestrahlung. Leipzig
86. Einstein A (1905) Über einen die Erzeugung und Verwandlung des Lichtes betreffenden heuristischen Gesichtspunkt. Annalen der Physik 322(6):132–148. ISSN:1521-3889. https://doi.org/10.1002/andp.19053220607
87. Bohr N (1913) On the constitution of atoms and molecules. Phil Mag Ser 6(26):1–24. https://doi.org/10.1080/14786441308634955
88. Bohr N (1913) On the constitution of atoms and molecules. 2. Systems containing only a single nucleus. Phil Mag Ser 6(26):476. https://doi.org/10.1080/14786441308634993
89. Olaf N, Markus A, Anton Z (2003) Quantum interference experiments with large molecules. Am J Phys 71(4):319–325. https://doi.org/10.1119/1.1531580

90. Donati O, Missiroli GP, Pozzi G (1973) An experiment on electron interference. Am J Phys 41(5):639–644. https://doi.org/10.1119/1.1987321
91. Schrodinger E (1926) Quantisierung als eigenwertproblem. Ann Phys 384(4):361–376. [Ann Phys 79(Ser IV), 361 (1926)]. https://doi.org/10.1002/andp.19263840404
92. Schrodinger E (1926) Quantisierung als eigenwertproblem. Ann Phys 385(13):437–490. [Ann Phys 80(Ser IV):437 (1926)]. https://doi.org/10.1002/andp.19263851302
93. Schrodinger E (1926) Der stetige Ubergang von der Mikro- zur Makromechanik. Naturwiss 14:664–666. https://doi.org/10.1007/BF01507634
94. Schrödinger E (1926) Quantisierung als eigenwertproblem. Ann Phys 384(6):489–527. [Ann Phys 79(Ser IV):489 (1926)]. https://doi.org/10.1002/andp.19263840602
95. Schrödinger E (1926) Quantisierung als eigenwertproblem. Ann Phys 386(18): 109–139. [Ann Phys 81(Ser IV), 109 (1926)]. https://doi.org/10.1002/andp.19263861802
96. Davisson CJ, Germer LH (1928) Reflection of electrons by a crystal of nickel. Proc Natl Acad Sci 14(4):317–322
97. Tanamura Dr (2018) Double-slit experiment. https://commons.wikimedia.org/wiki/File: Doubleslit_experiment_results_Tanamura_2.jpg. Accessed 13 Jan 2018
98. Gerlach W, Stern O (1922) Der experimentelle Nachweis der Richtungsquantelung im Magnetfeld. Zeitschrift für Physik 9(1):349–352
99. Gerlach W, Stern O (1922) Das magnetische Moment des Silberatoms. Zeitschrift für Physik A Hadrons and Nuclei 9(1):353–355
100. Gerlach W, Stern O (1922) Der experimentelle Nachweis des magnetischen Moments des Silberatoms. Zeitschrift für Physik A Hadrons and Nuclei 8(1):110–111
101. Knott T (2017) Stern-Gerlach experiment. https://commons.wikimedia.org/wiki/File:Stern-Gerlach_experiment_svg.svg. Accessed 21 Dec 2017
102. Dirac PAM (1928) The quantum theory of the electron. Proc R Soc Lond A117:610–624. https://doi.org/10.1098/rspa.1928.0023
103. Dirac PAM (1930) A theory of electrons and protons. Proc R Soc Lond A 126:360. https://doi.org/10.1098/rspa.1930.0013
104. Anderson CD (1933) The positive electron. Phys Rev 43:491–494. https://doi.org/10.1103/PhysRev.43.491
105. Fierz M, Pauli W (1939) On relativistic wave equations for particles of arbitrary spin in an electromagnetic field. Proc R Soc Lond A173:211–232. https://doi.org/10.1098/rspa.1939.0140
106. Chadwick J (1932) Possible existence of a neutron. Nature 129(3252):312
107. Breidenbach M et al (1969) Observed behavior of highly inelastic electron-proton scattering. Phys Rev Lett 23:935–939. https://doi.org/10.1103/PhysRevLett.23.935
108. Wok J (2017) Pseudorapidity. https://commons.wikimedia.org/wiki/File:Pseudorapidity2.png. Accessed 10 Sep 2017
109. Kühne S (2018) A mercator projection map with Tissot's indicatrices. https://en.wikipedia.org/wiki/File:Tissot_mercator.png. Accessed 08 Jan 2018

Chapter 3
The Large Hadron Collider and the Compact Muon Solenoid

The CERN[1] is an international centre for experimental physics [1]. Originally founded by twelve European countries in 1952 to associate their research programmes, many other countries from all around the world have now joined European Organisation for Nuclear Research (CERN) at various levels, and other topics of research have come up. CERN is nowadays one of the leading centres for research in nuclear and particle physics, for an internal budget of around one billion euros a year.

CERN hosts the largest complex of particle accelerators in the world (illustrated with the diagram in Fig. 3.1), and is therefore a unique place for performing all kinds of experiments involving high-energy beams, from particle physics to meteorology. In particular, it hosts the Large Hadron Collider (LHC), where beams are scattered head on at extremely high energy densities. Such a configuration implies the building of very large and complex detectors, such as the Compact Muon Solenoid (CMS). The present chapter is devoted to their descriptions.

In this section, we give a description of LHC and CMS, emphasising on aspects that will matter in the physics analysis.

3.1 Large Hadron Collider

The LHC project was approved in 1992 by the CERN Council, with the intention of succeeding the Large Electron Positron Collider (LEP), in the same tunnel and in the physics programme, especially in the search of the after the three physicists Brout, Englert and Higgs (BEH) boson [3, 4].

---

[1]The acronym stands for the original name in French of the organisation *Conseil Européen pour la Recherche Nucléaire*, and has been kept for the proximity with the Germanic root *kern*, meaning "nucleus".

© Springer Nature Switzerland AG 2019
P. L. S. Connor, *Inclusive b Jet Production in Proton-Proton Collisions*, Springer Theses,
https://doi.org/10.1007/978-3-030-34383-5_3

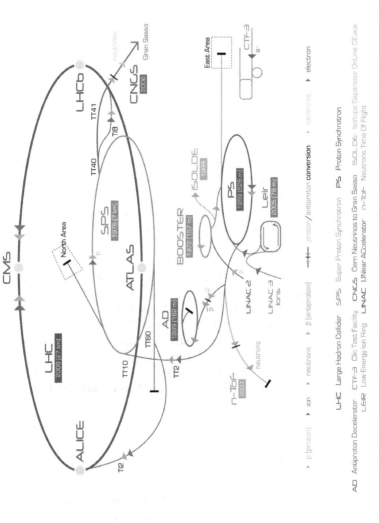

Fig. 3.1 CERN accelerator complex [2]. Protons are extracted from a simple bottle of hydrogen based at the LINAC 2. They are collected in the Booster that accelerate them from 50 MeV to 1.4 GeV, then to the PS that speeds them up to 26 GeV and separates the beams in smaller bunches, then to the SPS that makes them go to 450 GeV. And eventually, they can go to the LHC, whose nominal energy per beam in 2016 is of 6.5 TeV. Lead nuclei can be as well accelerated, but start from the LINAC 3 and go through the LEIR. Figure reproduced with permission from CERN

In this section, after a global overview of the project, we give the basic principles of particle acceleration and apply them to the LHC.

3.1.1 Overview

As can be seen from Fig. 3.1, the LHC is at the end of a long chain of accelerators, providing beams of protons. It is built underground to reduce the impact of cosmic rays in measurements. As of 2016, each beam can carry up to 6.5 TeV of energy in the centre-of-mass system of the colliding particles, highest energy in the world. This opens up new regions of the phase space to perform precision measurements—as in the case of the present thesis—or searches—like the search of the *Higgs boson*.

Especially, such a high energy allows the production of *b* jets up to transverse momenta of order from $\mathcal{O}(10\,\text{GeV})$ up to $\mathcal{O}(1\,\text{TeV})$. Figure 3.2 shows how LHC enlarges the phase space with respect to the HERA and Tevatron experiments.

LHC stands for Large Hadron Collider:

Collider   As its name suggests, the LHC is a *collider*, i.e. it accelerates hadrons in opposite directions and makes them collide in flight. Collider experiments are to be opposed to *fixed-target* experiments, where particles are all accelerated in the same direction on a fixed target. Although easier to set up and although providing higher luminosity, the energy in the centre-of-

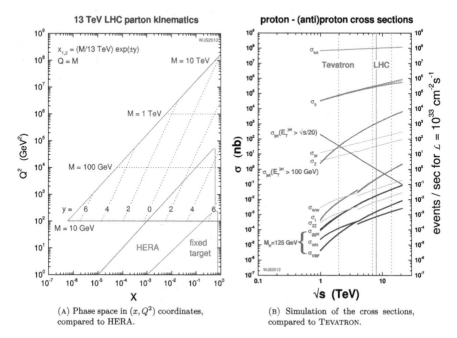

(A) Phase space in (x, Q^2) coordinates, compared to HERA.

(B) Simulation of the cross sections, compared to TEVATRON.

Fig. 3.2 Comparisons of the accessible phase space at LHC compared with previous experiments [5]

mass system grows as the square root $\sqrt{E}$ of the beam energy E of the incident proton, while it grows as $\sqrt{E_1 E_2}$, the square root of the product of the energies $E_{1,2}$ of the colliding beams. Therefore, colliders are more suited to investigate region of the phase space of higher energy.

Hadron   Furthermore, the LHC collides hadrons. In practice, these are either protons or lead nuclei:

- Protons, since the *proton* is the only stable and charged hadron that can be found in nature, allowing the study of fundamental parameters of the Standard Model (SM) or searches Beyond the SM (BSM). LHC currently provides proton collisions at 13 TeV in the centre-of-mass frame.
- Lead nuclei, since the *lead* is the heaviest stable nucleus (around 208 times the mass of the proton), allowing the study of *quark-gluon plasma*. LHC currently provides collisions at around 1150 TeV in the centre-of-mass frame, corresponding to around 2.75 TeV per nucleon inside of the lead nucleus.

In this thesis, we shall only discuss proton-proton collisions.

Large   Finally, the LHC is a circular collider, i.e. it is made of two superimposed rings in which protons circulate in respectively opposite directions. *Circular colliders* are to be opposed to *linear colliders*.

- In a linear (circular) collider, the accelerator and the bunches can be only once (many times). Yet reusing the same beams may save energy and time.
- However, circular colliders are affected by the *synchrotron radiation*[2] [6], i.e. part of the energy that is dedicated to the acceleration of particles is lost in the form of light rays. At each revolution, the loss of energy ΔE the goes as follows:

$$\Delta E = \frac{1}{3} \frac{e^2}{\epsilon_0} \left(\frac{E}{R(mc^2)} \right)^4 \tag{3.1}$$

where
- e is the electric elementary charge,
- ϵ_0 is the permittivity,
- E stands for the energy
- and R stands for the radius of the ring.

For protons at $E = 6.5$ TeV, $\Delta E \approx 0.01$ MeV. In addition, from this formula, one can also see the advantage of using protons instead of electrons: since protons are much heavier, the radiation is much more suppressed with a factor of $(m_e/m_p)^4 \approx 10^{-13}$.

---

[2]Note that in general, any charge that has a non-uniform movement should radiate according to the Maxwell equations; however the radiation in case of a linear collider are usually negligible.

Fig. 3.3 Surroundings of CERN and Geneva with the Lake Léman, highlighting the position of LHC underground (yellow line) and SPS (light blue line) [7]. The two campuses of CERN can be seen (Meyrin and Prévessin), as well as the four big experiments (CMS, ATLAS, LHCb and ALICE). It can also be compared to the airport to better appreciate its size. Figure reproduced with permission from CERN

The circumference of LHC is of 26.659 km, the *large* radius[3] of around 4.3 km allowing to limit the synchrotron radiation; the size of the ring can be compared to the size of the surroundings in Fig. 3.3.

The LHC machine is in constant development. Since the start of data taking in 2008, it has already undergone two *runs*. The 2016 data that we shall analyse later belongs to the second run, a.k.a. LHC Run-II. Further upgrades are intended in order to reach higher beam energies and luminosities for later data-taking periods.

3.1.2 Principles of Acceleration

Some general principles of acceleration are here given [1, 8].

The techniques used to accelerate particles are relativistic but non-quantum-mechanical. Particles are only manipulated through classical electromagnetic interactions.

---

[3]Note that the LHC is not perfectly circular, some section being linear.

From the extraction from a bottle of gaseous hydrogen to the collisions, four main aspects related to proton acceleration may be distinguished:

acceleration   Given $F = qE$, the main technique in use relies on *radiofrequency cavities* (or *resonance cavities*): cavities provide an alternating electric field $E(t)$, where charges will pass successively. In order to be only sensitive to acceleration phases of the electric fields, the charges stay in cavities only during a half period; during the other half periods, charges travel through tubes with no ambient electric field. The working principle is illustrated in Fig. 3.4. The length of the tubes and the frequency of the alternating electric field depend on the configuration:

- At the Linear Accelerator 2 (LINAC 2), since the protons start from rest, the first tubes are of variable size. Then, since the velocity gets closer and closer to the speed of light, the length of last tubes can be constant.
- However, in the case of a circular accelerator made of a ring, such at the LHC, since the same cavities are reused several times, the frequency needs to be varied. An accelerator relying on this principle is called *synchrotron*, such as most of the circular accelerators nowadays.[4]

clustering and bunch splitting   Particles in the beams are not all synchronous, i.e. they are not all perfectly synchronised with the electric fields in the resonance cavities. Early (late) particles will be more (less) accelerated, resulting in longitudinal oscillations of particles.

steering   Dipole magnets are in principle sufficient to steer particles. At LHC, the technical difficulty lies in getting strong enough magnets to steer the beams, requiring 8.33 T of magnitude. This has been made possible by the superconducting magnet technology.

focusing   Focusing the beams is required for several reasons:

- to keep same-charge particles confined in the beam;
- to control the transverse oscillation of the protons in the beams (i.e. protons should stay inside of the pipe);
- to focus strongly the two beams shortly before the collision so as to increase the cross section (see Fig. 3.5).

This may be achieved thanks to pairs of *quadrupole magnets*: one quadrupole magnet focusing in one transverse direction while defocusing in the other, two quadrupole magnets with alternating poles will act as consecutive convergent and divergent lenses in optics.

In addition, all operations on the beams require an very deep vacuum inside of the pipe, in order to avoid collisions with particles in the medium.

Given these elements, and given the diagram of the complex of accelerators at CERN in Fig. 3.1, we can now detail the different phases of acceleration of particles and beams of particles up to the collision at LHC:

---

[4]By opposition, the first circular accelerators were no synchrotrons but *cyclotrons*, made of a disk rather than a ring, with constant frequency but variable curvature of particles.

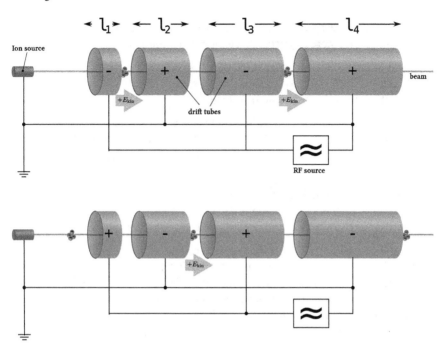

Fig. 3.4 Illustration of the working of radiofrequency cavities [9]. Cavities and tubes succeed to one another. Each cavity provides an alternating electric field. Here, the frequency is taken as constant and the length of the tube is adapted such that charges get only accelerated by the successive cavities according to $L = v \times \frac{T}{2}$

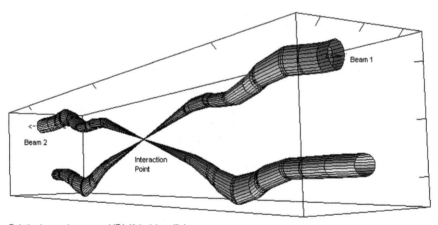

Relative beam sizes around IP1 (Atlas) in collision

Fig. 3.5 Increase of the density of the beams while approaching the IP1 (ATLAS) [10]. Note that such a focus could be achieved continuously, since it comes together with an increase of the amplitude of oscillation of the particles inside of the beam. Figure reproduced with permission from CERN

LINAC 2   The Linear Accelerator 2 is an injector, i.e. it is only designed to provide protons to other, more powerful accelerators. Its rôles are:

– extraction of protons from bottle of hydrogen,
– acceleration from rest to 50 MeV (for a speed around 0.3 c),
– continuous beam production.

Booster   The Proton Synchrotron Booster [11] is made of four superimposed rings for different functions; each ring may accelerate one or two bunches from the injection of protons from the LINAC 2. In its rôle of injector for LHC,[5] its functions are:

– acceleration from 50 MeV to 1.4 GeV,
– accumulation of protons from LINAC 2 (or LINAC 3 for heavy nuclei),
– production of six or eight bunches (different schemes are possible).

PS   The Proton Synchrotron is the oldest accelerator still in use at CERN. It is also the very first synchrotron at CERN, with a length of 628 m. After a period dedicated to physics in the fifties, it underwent several upgrades and is now only used as an acceleration facility of various kinds of particles to different machines. In the chain of proton acceleration to LHC, getting bunches from the Booster, it is used for:

– acceleration from 1.4 to 24 GeV,
– bunch splitting (typically from 6 to 72 bunches [12]).

SPS   The Super Proton Synchrotron was the successor of the PS, with nearly 7 km of circumference, and has also been recycled as an accelerating facility:

– acceleration from 24 to 450 GeV.

LHC   Finally, protons reach the LHC, which will take care of the following:

– acceleration from 450 GeV to (as of 2016) 6.5 TeV,
– collision of beams at CMS and other experiments.

Once the beams have reached their cruising speed, their bunches measure approximately as long as a knitting needle and follow one another with a distance around 7.5 m. Collisions may take place at the different interaction points (IPs), corresponding to the different experiments.

3.1.3   Collisions

In normal running conditions, crossing bunches typically provide a pile-up of a few tens of collisions. The term *pile-up* generally refers to the fact that more than one

---

[5] It is interesting to keep in mind that most protons accelerated at CERN are not intended to be delivered at LHC. The Booster has various working modes and deliver beams of different flavours. In fact, the experience requiring most of the protons at CERN is ISOLDE.

pp collision may occur at the same time in a single bunch crossing[6]; it will be discussed in more detail in Sect. 7.2 while performing the analysis. The pile-up is the price to pay to increase of the rate of interesting processing without degrading too much the quality of the recorded data.

The *total cross section*[7] σ_{tot} at LHC at $\sqrt{s} = 13$ TeV is around 100 mb [14, 15]. It can be distinguished between the *elastic cross section* (around 30 mb) and *inelastic cross section* (around 70 mb).

Since the total cross section defines the collision rate, one defines the *luminosity* of an accelerator to relate it to the number of delivered collisions:

$$N_{tot} = \sigma_{tot} \times L \tag{3.2}$$

It characterises the performance of the accelerator. This formula can be applied to any process, i.e. to any cross section:

$$N = \sigma \times L_{int} \tag{3.3}$$

The luminosity itself is independent of the process, and is only related to the amount of delivered collisions.

In practice, one defines the *instantaneous luminosity* and the *integrated luminosity*.

Circular colliders with identical beams deliver a luminosity according to the following formula:

$$L_{inst} = fn\frac{N^2}{A} \tag{3.4}$$

where

- f is the revolution frequency (around 11 kHz),
- n the number of *bunches* in the ring (around 2800),
- N the average number of particles in one bunch (around 10^{11})
- and A the *cross-sectional area*[8] of the beams (around 10^{-5} cm^2).

---

[6]"Pile-up occurs when the readout of a particle detector includes information from more than one event. This specifically refers to cases where there are other (background) collisions somewhere within a timing window around the signal collision. [13]".

[7]Note that in principle, given the Coulomb interaction, the total cross section is divergent. The total cross section accounts therefore for nuclear interactions.

[8]The cross-sectional area can be further described as follows:

$$A = \pi\frac{\epsilon_n\beta^*}{\gamma}F \tag{3.5}$$

where

- ϵ_n is the *normalised transverse beam emittance*, which is representative of the intrinsic dispersion of the beam depending only on the initial conditions,
- β^* is the *amplitude function* at the IP, which is representative of the power of focusing of the magnets,
- γ is the relativistic factor,

This quantity is characteristic of the production of collisions by the machine per unit of time and is called the *instantaneous luminosity*. The smallest unit of time on which the luminosity is measured, is called Lumi Section (LS), corresponding to 2^{20} orbits (around 93 s) [13]. For LHC Run-II, the instantaneous and integrated luminosities are shown in Fig. 3.6.

In the regard of a physics analysis, only the *integrated luminosity* is given, since only the total amount of data matters to describe the statistics [17]. The total luminosity of the data with which we shall work is:

$$L_{int} = \int L_{inst} \, dt = 35.2 \pm 0.8 \, \text{fb}^{-1} \qquad (3.6)$$

Note that this value does not match with the recorded luminosity of 37.76 fb^{-1} from Fig. 3.6: indeed, while and after recording, each run is carefully monitored and only *certified* runs can be considered in physics analyses. Moreover, some special runs are dedicated to specific analyses, like the measurement of the Minimum Bias (MB) cross section, which requires low-pile-up conditions.

Since the cross-sectional area A varies with time (see Fig. 3.7), the beam luminosity decreases with time. At the LHC, a *run*[9] corresponds to the time that a beam stays in the pipe; runs are of variable luminosities and durations, but can reach a total luminosity of $\mathcal{O}(100 \, \text{pb}^{-1})$ and last twelve hours. The smallest unit of count for the instantaneous luminosity is called LS, corresponding to $\sim 2 \times 10^{20}$ orbits (around 93 s) [13].

3.2 Compact Muon Solenoid

The CMS detector [19] is one of the four main[10] detectors at LHC: CMS, ATLAS [24] (A Toroidal LHC ApparatuS), LHCb [25] (LHC beauty) and ALICE [26] (A Large Ion Collider Experiment). Like ATLAS, it is a general-purpose detector, i.e. it has been built in order to investigate various aspects of High-Energy Physics (HEP) (thus corroborating findings); it is to be opposed to LHCb—which was built mainly to investigate heavy-flavour physics and CP violation—or to ALICE—which

---

– and F is correction coming from the fact the beams do not collide exactly head on, but with a small angle of 300 μrad.

[9]The utilisation of the word "run" may be quite confusing, since it is used at three different scales: 1. it stands for long periods of data taking such at LHC Run-I and LHC Run-II (usually written with a capital letter and followed by a Roman number); 2. it stands for variable periods of data taking corresponding to the equivalent of a few weeks of data taking (in this context, also called *era*, and is usually followed by the year of data taking plus some capital letter(s)); 3. finally, as in this section, it corresponds to the data taken from a single pair of beams (corresponding to a 6-digit number).

[10]There are a few additional experiments of smaller size: TOTEM [20], LHCf [21], MoEDAL [22], CASTOR [23], …

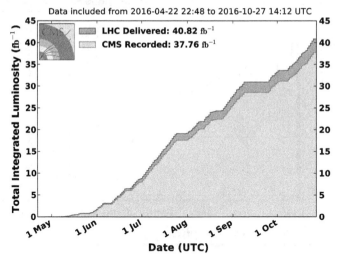

(A) "Cumulative offline luminosity versus day delivered to (blue), and recorded by CMS (orange) during stable beams and for pp collisions at 13-TeV centre-of-mass energy in 2016. The delivered luminosity accounts for the luminosity delivered from the start of stable beams until the LHC requests CMS to turn off the sensitive detectors to allow a beam dump or beam studies. Given is the luminosity as determined from counting rates measured by the luminosity detectors after offline validation."

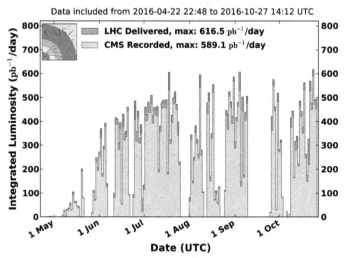

(B) "Same as the above, but not cumulative."

Fig. 3.6 Integrated luminosity at CMS and LHC in 2016 for proton-proton collisions at $\sqrt{s} = 13\,\text{TeV}$ (plots and captions taken from [16])

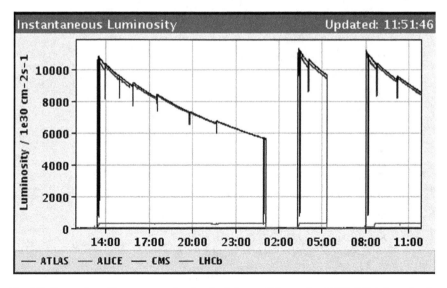

Fig. 3.7 Screenshot of instantaneous luminosity (taken on 8 September 2017) [18] delivered to the different experiments at LHC. A beam can last up to twelve hours; afterward, the luminosity gets divided by two with respect to its original value, and the beams are dumped and replaced

Fig. 3.8 Logo of the CMS collaboration. The different layers represent the subdetectors and the four tracks represent muons crossing the detector

was built mainly to investigate heavy-ion collisions and in particular the *quark-gluon plasma* state (see Sect. 2.2.5.1).

The acronym of CMS stands for Compact Muon Solenoid (Fig. 3.8):

Fig. 3.9 Panoramic picture of the CMS detector, taken in January 2017. During this shutdown period, the detector was opened in order to check and repair some components; one can see the beam pipe in the middle, and on the RHS (LHS) the *forward* (*barrel*) parts of the detector

Compact means that most of the devices are placed inside the magnet, the trackers and the calorimeters except for the muon chambers (building such a large magnet is a technical achievement);

Muon means that the detector has been designed to be very sensitive to the muons, as they are involved in some important processes;

Solenoid stands by opposition to *toroidal*, which is one of the two possible geometries for the magnetic field to be parallel to the beams in order to act only on the produced particles.[11]

As the name suggests, CMS is designed for particles scatterings in the *c*entre-of-*m*ass *s*ystem (even though it is not the case when studying proton-lead collisions).

The technical proposal [27] was written in 1994 and the construction began in 1997. A picture of the CMS detector is shown in Fig. 3.9 and a few key figures are given in Table 3.1.

The design of the detector was developed according to the following points [19]:

1. good muon identification up to 1 TeV and good resolution on the mass of dimuons (see muon chambers in Sect. 3.2.2.3);
2. good tracking and vertexing,[12] especially regarding the tagging of b jets and tauons (see tracker system in Sect. 3.2.2.1);
3. good resolution on electromagnetic energy and on the mass of diphotons and dielectrons (see ECAL in Sect. 3.2.2.2);
4. good resolution on Missing Transverse Energy (MET) and on the mass of dijets (see HCAL in Sect. 3.2.2.2).

All these points are crucial to the analysis presented in this thesis:

---

[11] Alternatively, ATLAS was designed with a toroidal configuration.

[12] In the context of reconstruction, a vertex corresponds to the crossing of several tracks, usually corresponding to the decay of a particle into several other particles.

Table 3.1 A few key figures comparing CMS to an A380 airplane (see also [1])

(A) Comparison to an A380 airplane

CMS	A380
100 m underground	1.25 × of the length
21 m long	0.3 × fuselage size
15 m diameter	Same height
14000 t	28 airplanes
5000 persons	10 airplanes

(B) CMS in numbers, at the time of writing the thesis

Category	Number
Active people	5250
– Staff physicists	1963
– Physics doctoral students	922
– Undergraduates	994
– Engineers	995
– Technicians	279
– Other	97
Institutes	198
Countries & regions	45

- the identification of secondary vertices, which may indicate the decay of a B hadron, mostly relies on the tracker;
- the identification of electrons and muons plays a rôle in b tagging, since a non-isolated soft lepton may also be the sign of the decay of a B hadron;
- the electromagnetic and hadronic calorimeters are essential in the reconstruction of jets.

In this thesis, we shall focus on issues related to heavy flavour rather than jet reconstruction.

In the next subsections, we give some general principles of reconstruction; then we explain how these are applied at CMS by reviewing its different subdetectors; eventually, we describe how (and when) an event is reconstructed.

3.2.1   General Principles of Detection and Reconstruction

We give here general principles of particle detection [28] and their specific application at CMS for the event reconstruction.

The interaction of particles with media are considered as classical. An event is described by the list of the particles, their four-momenta (see Fig. 2.13a for the conventional coordinate system) of the different processes that have happened at the IP.

Most of the particles created at the IP do not live long enough to be directly detected. As already addressed in Sect. 2.3.2.1, only the proton and the neutron can be considered as stable hadrons, whereas all other hadrons are expected to decay; the decay lengths of hadrons have been summarised in Table 2.5). Eventually, in the detector, one can find the following particles (and their antiparticles if relevant):

- photons (γ),
- electrons (e),
- muons (μ),
- charged pions ($\pi^{\pm}$),
- charged kaons ($K^{\pm}$),
- neutral kaons (K^{0}),
- protons (p),
- neutrons (n).

Neutrinos do not decay but interact too weakly to be detected in collider experiments; their presence will be estimated using the missing energy.

Those particles' interactions with media are very well-known:

Photons

- *photoelectric effect*,
- *Compton effect*
- and *pair production* by interaction with nuclei.

In principle, all types of interaction can take place, but at the energy scale of incoming photons in an event at CMS, the pair production is more significant, down to $\sim$10 MeV.

Electrons and muons

- *Bremsstrahlung*,
- *ionisation*
- and *multiple scattering*.

For the detection of electrons (muons), the Bremsstrahlung (ionisation) is the most significant interaction with the media down to $\sim$10 MeV.

Protons, charged pions and charged kaons Regarding the electromagnetic interactions, they are similar to the muons. However, they mainly interact by nuclear interactions.

Neutrons, neutral kaons Only have nuclear interactions.

In addition, mesons may also decay in flight: charged pions may decay weakly into muons for instance.

In practice, except for muons that continue through the magnet and through the *muon chambers* and neutrinos that interact too weakly to be seen, all particles produced at the IP go through the *tracker*, and should be stopped in one of the *calorimeters*:

tracker The aim is to reconstruct the trajectory of all charged particles coming from the IP. Their trajectories are curved thanks to the magnetic field, and the curvature of a trajectory can be related to its transverse momentum, according to the following relation:

$$\frac{p_T}{\text{GeV}} = 0.3 \frac{B}{\text{T}} \frac{\rho}{\text{m}} \tag{3.7}$$

where ρ is the curvature radius of the charged particle; in 3-dimensional space, this corresponds to a helical motion.[13]

calorimeter   The principle of a calorimeter is to stop all incident particles (except muons and neutrinos) and measure their energy deposits. If possible, a deposit will be associated to one of the charged particles seen by the tracker at reconstruction (in the tracker acceptance).

muon chambers   Muon chambers are a kind of external tracker designed especially for muons, the largest component of the CMS detector.

Given these elements, we can now give a description of the components of the CMS detector.

3.2.2   Application at CMS

A transversal view of the CMS detector can be seen in Fig. 3.10, and the pseudorapidity coverage is given in Table 3.2 according to the different subdetectors. From now on, the discussion will be specific to CMS, especially focusing on the parts of the detector that need to be covered to apply b-tagging techniques. Since b-tagging requires all components of the detector, it is limited to the acceptance of muon chambers, i.e. $|\eta| < 2.4$.

---

[13]In general, the motion of a particle in an electromagnetic field is given by *Lorentz force*:

$$\mathbf{F} = q\,(\mathbf{E} + \mathbf{v} \times \mathbf{B}) \tag{3.8}$$

Here,

–   the electric field $\mathbf{E}$ is null,
–   the magnetic field $\mathbf{B}$ is constant and parallel to the axis,
–   and the motion is supposed to be circular under the assumption that the crossing of the tracker is negligible.

Therefore, the transverse momentum can be obtained as follows:

$$m\frac{v_T^2}{\rho} = q v_T B \quad \rightarrow \quad p_T = q B \rho \tag{3.9}$$

Finally, the modulus of the momentum is simply obtained with trigonometry.

$$p = \frac{p_T}{\sin \theta} \tag{3.10}$$

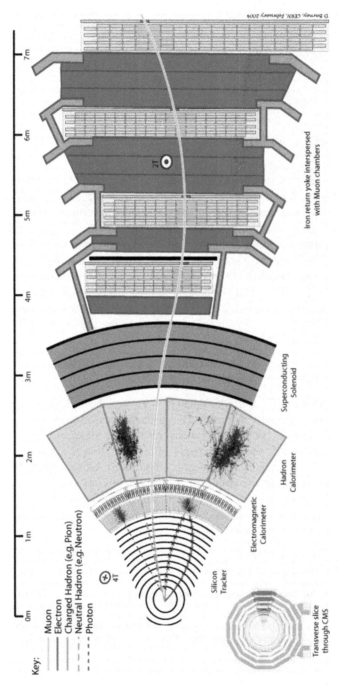

Fig. 3.10 Diagram of the barrel of the detector (end-caps have a similar structure) [29]. The ideal trajectories of the different types of particles have been drawn (see colour code in top right). A particle coming from the IP first goes through the tracker; it is seen only if it is charged. Then it goes through the electromagnetic calorimeters, which should only stop photons and electrons. If it has not been absorbed yet, the particle continues through the hadronic calorimeter, which should stop all hadrons. Only muons or neutrinos should still survive these steps, going then through the magnet and the steel structure, and through the muon chambers. There, muons will be detected while neutrinos will continue

Table 3.2 Pseudorapidity coverage of the different subdetectors at CMS

Subdetector	Coverage
Tracker	$|\eta| < 2.5$
ECAL	$|\eta| < 3.0$
HCAL	$|\eta| < 5.2$
Muon chambers	$|\eta| < 2.4$

3.2.2.1 Tracking and Vertexing

Overview. As already mentioned, the tracker's rôle consists in reconstructing the tracks of outgoing charged particles; especially, thanks to the magnetic field, it allows to determine their transverse momenta. At CMS, the tracker itself is made of *semi-conductor modules*, assembled in layers; the magnet is made of superconducting niobium-titanium, providing an homogeneous magnetic field of 3.8 T around the beam pipe. The interest of the semi-conductor technology is that the valence band of the electrons is just below the conduction band; the excitation of valence electrons makes them jump in the conduction band. Therefore, when a high-energetic charged particle goes through a module, a signal is induced, called a *hit*; combining hits, one can in principle reconstruct the tracks of all charged particles crossing the tracker, and consequently identify vertices.

Description. The semi-conductor modules are made of npn-doped junctions in silicon. At CMS, one distinguishes to types of modules:

silicon strip tracker   From $R = 55$ cm to $R = 110$ cm, the tracker consists of layers of doped semi-conductor detectors. The strips measure 25 cm $\times$ $180\,\mu$m and are arranged in stereo to get the two components of the coordinate.

pixel tracker   Closer to the beam pipe, from $R = 20$ cm to $R = 55$ cm, pixel cells are used in order to provide high accuracy. Pixel cells measure $100 \times 150\,\mu$m^2 to be as precise as the silicon strip tracker; the principles of working and detection are the same. The pixel cells must be made of materials resisting to the important radiations; however, significant degradations may be observed during the data-taking periods and continuous calibration is needed.

Pictures of pixel and strip modules are shown in Fig. 3.11. The pixel modules provide a finer resolution than the strip modules, as described in Table 3.3.

Track reconstruction. By combining tracks in successive layers, the tracking system allows to reconstruct tracks of charged particles; these can be later associated to Primary Vertices (PVs) or Secondary Vertices (SVs). At CMS, the tracking algorithm [30] is based on the *Kálmán filter* [31]; it is an iterative procedure, applied layer by layer, as shown in Fig. 3.12:

1. On the first layer, one can only measure the position; the momentum is arbitrarily set to 5 GeV, with a very large uncertainty.

(A) Two pixel modules. One sees here the read-out chips. The sensors are on the other side of the modules.

(B) A strip module. One can see the read-out chip on the left, and the two sensors (the reflecting surfaces) of the strip module.

Fig. 3.11 Silicon modules as of 2016 (pictures taken at DESY in February 2018)

Table 3.3 Description of the pixel and strip modules

	Units	Hit resolution
Pixel	1440	$9\,\mu m$
Strip	15 148	$20–60\,\mu m$

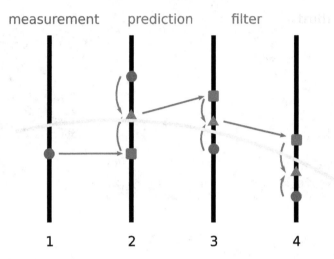

Fig. 3.12 Illustration of Kálmán filter in a simplistic case with 4-layer tracker: the iterative procedure is performed layer by layer. The layers are represented transversally in black; the true track is represented with the bended yellow curve. The position on the second layer is predicted from the position on the first layer (red arrow); then a compromise—the *filter* (green triangles)—is found (green arrows) between the measurement (blue points) and the prediction (red squares). The prediction for the position on the third layer is performed taking into account the kinematics from the two first layers, etc.

2. On the next layer, the position can be both a) determined experimentally and b) predicted by extrapolation from the previous layer(s) with the Equations of Motions. The *filtered position* is then determined as a compromise between the prediction and the measurement. The balance accounts for the respective uncertainties.
3. The track parameters are updated with the last filtered position, and the position on the next layer is predicted. Item 2 is repeated till the last layer.

Many refinements exist:

– The filter can be tuned for noisy environment, where many tracks have crossed the tracker. In particular, at CMS, a preselection of candidate tracks is considered before applying the reconstruction, in order to reduce the combinatorics.
– It can also account for inefficiencies (i.e. for missing hits) and for the thickness of the material (i.e. for multiple scattering).
– The filter can be applied twice to reduce sensitivity to the 5 GeV seed: once starting from the innermost layer, once starting from the outermost layer. It can also be

reiterated taking into account the vertex to which it belongs, once the vertices have been reconstructed.

At CMS, each track is typically reconstructed with 20 hits in the tracker, and is efficient down to 1 GeV.

Vertex reconstruction. Given the reconstructed tracks, one can reconstruct (PVs) and (SVs). In practice, in normal run conditions, a few tens of vertices are expected per bunch crossing. The reconstruction of vertices is another problem of pattern recognition. The adopted strategy at CMS is called Adaptive Vertex Reconstruction (AVR) [30]; given the complexity of the problem of vertexing, it can here only be outlined:

1. Based on the position with respect to the beam pipe (z), candidate vertices are proposed by the *deterministic annealing* technique [32]. The deterministic annealing technique treats the problem in analogy with statistical mechanics, defining a free energy and a temperature; starting from only one cluster of tracks, the temperature is gradually lowered, a phase transition corresponding then to the splitting of a cluster. The difficulty lies in finding the right balance between clustering a variable number of tracks in a variable number of vertices, avoiding as much as possible to split a genuine interaction into two clusters of tracks. In this work, only tracks with $z < 24$ cm have been considered.
2. Based on the 3D coordinates $(x, y$ and $z)$, an Adaptive Vertex Fitter (AVF) [33] is applied to determine the best estimates of the vertex configuration from all track parameters. It consists in a modified, more robust Kálmán filter, where all tracks are weighted according to their compatibility with the vertex; in particular, outliers are downweighted.

This method is used to determine all vertices. SVs are selected among all vertices with additional requirements, depending on the analysis; another vertex fitter, the Inclusive Vertex Fitter (IVF), is sometimes used [34].

Resolution. The resolution on the momentum (Eq. 3.10) is essentially limited by two factors [35]:

curvature The higher the transverse momentum, the straighter the curve. The tendency increases with the transverse momentum:

$$\frac{\sigma_p}{p} \propto p_T \tag{3.11}$$

multiple scattering The lower the momentum, the more sensitive to multiple scattering. The effect is almost constant with respect to transverse momentum but becomes significant when the curvature becomes smaller:

$$\frac{\sigma_p}{p} \propto \frac{1}{\sin\theta} \tag{3.12}$$

Alignment. The tracker needs to be aligned for two reasons:

1. The precision of mounting of the modules at assembly is lower than their hit resolution.
2. A systematic misalignment may introduce a systematic bias in the reconstruction, which may then lead in a systematic bias in any measurement.

Therefore the resolution can be improved by aligning the modules. Details on alignment procedure may be found in Appendix A.

Calibration. A high-voltage tension of a few hundreds Volts is used to control the sensitivity of the modules, and needs to be adjusted in order to get a uniform sensitivity throughout the whole volume of the tracker. Moreover, due to the high radiation environment during data taking, the sensitivity region of the modules reduces significantly with time; to counter-balance the ageing of the modules, the high-voltage tension needs to be readjusted regularly. Additional details on the calibration of the modules may also be found in Appendix A.

3.2.2.2 Calorimetry

Overview. While the principle of a tracker is to measure the momenta of the charged particles without significant interaction, the principle of a calorimeter is to measure the total energy by stopping completely charged and neutral particles (only muons and neutrinos are not affected). Incoming high-energetic particles provoke cascades of particles of smaller energy. At CMS, one distinguishes two calorimeters: the Electromagnetic CALorimeter (ECAL) and the Hadronic CALorimeter (HCAL); the former (latter) is more suited to the detection of photons and electrons (hadrons).

Description. The ECAL and HCAL [36] are (almost) entirely placed between the tracker system and the magnet, as shown in Fig. 3.13. However, they have different structures and are made of different materials:

ECAL   The ECAL is an *homogeneous calorimeter*, which means that the entire volume is used to collect the signal, situated between the tracking system and the HCAL. According to the region of the detector, one distinguishes the ECAL Barrel (EB) and the ECAL End-caps (EE); the geometry of the ECAL is given in Table 3.4a. Both regions are made of a lead tungstate crystals with photodetector glued onto the back. The high density ($8.3\,g/cm^3$, each crystal weigh around $1.5\,kg$) make the electrons and positrons radiate, and make the photons produce electron-positron pairs. The scintillator medium is transparent for photons of a certain wavelength that is subsequently collected by photodetectors. In addition to the EB and EE, a *pre-shower* detector is installed in $|\eta| < 0.9$ and $1.65 < |\eta| < 2.61$ in order to help distinguishing pions and photons.

HCAL   The HCAL exploits the nuclear interactions of the hadrons to produce lighter hadrons and photons. It consists in a *sampling calorimeter*, structured in *towers* made of alternate layers of absorbers and scintillators, the former (latter)

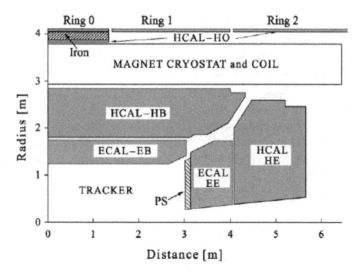

Fig. 3.13 Diagram showing the longitudinal view of the calorimeters and of the magnet [37]. The blue (green) region stands for the ECAL (HCAL). Only the HF is not represented. The rings correspond to the mechanical structures holding the whole detector

Table 3.4 Geometry of the calorimeters at CMS [36]. The numbers may sometimes be rounded up for more readability

(A) ECAL

	Coverage			Crystals	
	Rapidity ($|\eta|$)	Radius (r/m)	Position ($|z|$/m)	Dimensions (V/cm^3)	Number
EB	0−1.48	1.24−1.75	0−3.0	2.2 × 2.2 × 23	61 200
EE	1.48−3.0	0.32−1.71	3.2−3.9	2.86 × 2.86 × 22	2 × 7324

(B) HCAL

	Coverage			Towers	
	Rapidity ($|\eta|$)	Radius (r/m)	Position ($|z|$/m)	Coverage ($\Delta\eta \times \Delta\phi$)	Number
HB	0−1.4	1.8−2.9	0−4.0	0.087 × 10°	2304
HO	0−1.26	3.85−4.1	0−2.5	0.087 × 30°	
HE	1.4−3.0	0.45−2.9	3.9−5.7	0.09−0.35 × 5−10°	2304
HF	3.0−5.2	0.15−1.3	11.2−12.85	0.1−0.3 × 10−20°	1800

being used to slow down the hadrons (collect the produced photons). The HCAL consists in four regions: the HCAL Barrel (HB), the HCAL End-caps (HE), the HCAL Forward (HF) and the HCAL Outer (HO); the geometry of the HCAL is given in Table 3.4b. It is situated outside of the ECAL and inside of the magnet, except the HO, placed outside of the magnet together with the muon chambers.

The high granularity of both calorimeters is crucial to perform a fine description of the jet substructures, and to determine the contamination from the pile-up activity. It also implies that the alignment of the crystals with one another and with the tracker is crucial; however, this topic will not be treated here.

Reconstruction. The energy deposits are collected with low-energy photons in scintillators, using Compton effect. Deposits in the calorimeter will be compared to tracks in the tracker and in the muon chambers, distinguishing thus deposits from charged particles and deposits from neutral particles. Jets are then defined by clustering the particles,[14] following one of the algorithms described in Sect. 2.3.2.2; this is illustrated in Fig. 3.14.

Resolution. In general, the energy resolution is described by the following formula [38]:

$$\left(\frac{\sigma_E}{E}\right)^2 = \left(\frac{S}{\sqrt{E}}\right)^2 + \left(\frac{N}{E}\right)^2 + C^2 \tag{3.13}$$

where

- S is the stochastic term, accounting for statistical fluctuations in the cascade detection;
- N is the electric noise,
- C is a constant contribution, standing for miscalibrations.

The stochastic term is more (less) important for the HCAL (ECAL) since it is a sampling (homogeneous) calorimeter, but its effect becomes less relevant at high energy. In addition, given the structure of the CMS detector and the different rapidity coverages of the different subdetectors, the energy resolution in the calorimeters also depends on the pseudorapidity; details may be found in [39]. Details related to the jet energy calibration will be addressed while performing the analysis in Chap. 7.

3.2.2.3 Muon Detection

Overview. There are three types of *muon chambers*. They all are gaseous detectors: the principle is to fill a volume with a gas and an electric field, so that an incoming particle may ionise the gas and produce an electric signal. Different types of gaseous detectors exist, according to the handling of the electric signal.

Description. In the case of the muon chambers, one may distinguish three types of detectors:

CSC   Cathode Strip Chambers are based on the principle of multi-wire proportional chambers. They are made of arrays of anode wires and cathode strips, arranged

---

[14]Note that it is also possible to define "calo jets" uniquely from the energy deposits in the calorimeters. However, since this does not allow to apply *b*-tagging techniques based on secondary vertices, this has not been considered in the present work.

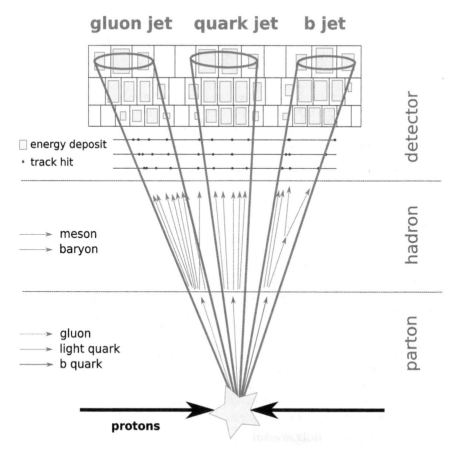

Fig. 3.14 Sketch of the jet reconstruction: first the collision takes place and partons are produced in collimated regions, and undergo hadronisation; after hadronisation, the hadrons traverse the tracker and the calorimeter

perpendicularly within a gas volume. The gas is ionised by the passage of a muon, its position being determined by the intensity of currents induced on the strips. The 540 CSCs are arranged in six layers in the end-caps.

DT   Drift Tubes are situated in the barrel ($|\eta| < 1.6$). Their dimensions are 4 cm × 2 m × 2.5 m, in which an anode wire is stretched; they are arranged in layers perpendicularly to the muon trajectories. When a muon goes through a DT, the gas is ionised, and the motion of the released electrons and ions induces a signal. The delay of the signal is then used in order to determine the position of the muon. As a DT gives only one coordinate, 3 × 60 DTs are arranged in three layers: the first and the last layers are used to measure the perpendicular coordinates, whereas the middle one is used to measure the coordinate parallel to the beam pipe.

RPC   Resistive Plate Chambers are made of an anode and a cathode plates separated by a gas volume. The material that are used are highly resistive to intense electric fields: $E \simeq 50\,\mathrm{kV/cm}$. 480 (288) RPCs are arranged in four concentric cylinders (four disks) in the barrel, $|\eta| < 1.6$ (in each end-cap, $|\eta| > 1.6$). RPCs are also used to obtain an estimation of the momenta of the muons in real time; however, the spatial resolution is moderate.

The muon chambers are combined with the central tracker in the reconstruction of the trajectories of the muons; they improve both the identification and the determination of the transverse momenta. A procedure of alignment of the muon system with itself and with the tracker is also required, but is not discussed here.

Reconstruction. Roughly, the reconstruction of a muon is performed in different steps:

local   inside a muon chamber,
standalone   combination of all muon chambers
global   combination of the muon system and the tracker.

Refinements exists; for instance, muons are sometimes primarily reconstructed in the tracker and related to hits in the muons chambers if the hits in the muon chambers were not enough to determine the passage of a muon. The identification of muons is rather performant as they must have been seen in the muon chambers. Several levels of quality have been defined to describe the muon reconstruction.

tight   optimised for weak boson studies, with reconstruction from both the tracker and in two different muons chambers;
soft   optimised for b quark decays, with reconstruction in one CSC or in one DT;
loose   optimised for multiple-muons events, with reconstruction after the full combination of the tracker, the muon chambers and the calorimeter;

Muons are mainly mentioned for CMS has been especially designed to have a performant muon identification; however, muons do not play a crucial rôle in this thesis.

Resolution. The resolution follows the same principles as for standard tracking.

3.2.2.4   Global Event Reconstruction

In a nutshell, the reconstruction goes as follows:

– Muons are identified easily thanks to the muon chambers.
– Electrons and positrons are reconstructed thanks to the association of a curved track and a deposit of energy.
– Photons are identified only thanks to the energy deposits in the ECAL.
– Hadrons are associated to all the left deposits in the HCAL.
– Jets are clustered from collimated particles (see Fig. 2.3).
– Neutrinos are not reconstructed but their presence is estimated from the MET.

This technique of reconstruction is called Particle-Flow (PF) algorithm [40]. In practice, the PF algorithm is tuned for each experiment, according to the specificities.

Moreover, the algorithms of reconstruction of the whole event are usually more complex than the ones introduced in this chapter, accounting for magnetic effects, pile-up, superposition of the tracks coming from the same bunch crossing, decays in flight, detector inefficiencies, etc.

3.2.2.5 Trigger and Storage Systems

The LHC is designed to acquire data at a very high collision rate, so high as 40 MHz; in 2016, bunches follow one another every 25 ns. At each bunch crossing, tens of collisions take place—the *pile-up*—, and each collision may provide hundreds of particles. Given that one particle must be detected in several parts of the detectors that have to be combined to reconstruct its trajectory and determine its nature, the total amount of data produced at each second is equivalent to several thousands of Wikipedia encyclopedias. This represents a too high volume to be treated or even stored. Therefore, a *trigger system* is set up, selecting potentially interesting events [41]. In practice, a very few events are accepted: around one event for a few hundred thousands.

At CMS, the trigger system is composed of two levels:

L1   The Level 1 applies at the *hardware* system, only involving the calorimeters and the muon chambers for a stream of order of 100 kHz.

HLT   The High-Level Trigger corresponds to the *software* system and is itself organised in different sub-levels:

- the `Level 2` also relies on the information from the calorimeters and from the muon chambers.
- the `Level 2.5` combines the information at `Level 2` with tracks in the pixel tracker in order to determine the region of extrapolation in the strip tracker.
- finally the `Level 3` combines `Level 2.5` with the strip tracker.

The stream delivered by HLT is of order of 100 Hz.

The whole event is then registered, including all electronic signals in order to investigate possible biases or dysfunctions. The data is sent to different computing centres: first the `Tier 0` at CERN, then to other computing and storing centres called `Tier 1` and `Tier 2` based in participating institutes and universities. The data files will then be further investigated, or *certified*, in order to determine whether they are valid for physics analyses.

Summary

The particles have followed a long procedure of extraction and acceleration until they are grouped by bunches and ready to collide. Bunches cross, a few tens of protons collide, and as a result, many particles are produced in the beam pipe; most of them decay. The remaining ones cross the devices of the detector: first the particles go through the tracker, aiming at reconstructing the trajectories of the charged particles.

Secondly, the particles pass through the ECAL, which stops photons as well as positrons and electrons with electromagnetic interactions. Thirdly, the remaining particles go to the HCAL; the same procedure applies to the hadrons, but adapted to their nuclear interactions. Besides, the muons are the only particles left that can still be detected; this is achieved by the muon chambers.

With the L1 trigger, one can decide whether an event is interesting to keep or not only on the basis of the calorimeter and the muon chambers; if the L1 level triggers, the HLT refines the selection of calorimeters and reconstructs tracks. After the selection (a few tens per second at most) the data is sent to the computing centres of the CERN and its partners (`Tier 0, 1, 2`), and parsed in data files. After a careful check of data quality, most runs are certified for physics analyses.

References

1. CMS Collaboration. CMS public web site. http://cms.web.cern.ch. Accessed 09 Aug 2017
2. Lefèvre C (2008) The CERN accelerator complex. Complexe des accélérateurs du CERN. https://cds.cern.ch/record/1260465
3. CERN. LHC the guide. https://cds.cern.ch/record/2255762/files/CERN-Brochure-2017-002-Eng.pdf. Accessed 06 Oct 2018
4. Lyndon Evans and Philip Bryant (2008) LHC Machine. J Instrum 3(08):S08001. http://stacks.iop.org/1748-0221/3/i=08/a=S08001
5. WJ Stirling (2012) Private communication. http://www.hep.ph.ic.ac.uk/~wstirlin/plots/plots.html. Accessed 07 Jan 2018
6. Elder FR et al (1947) Radiation from electrons in a synchrotron. Phys Rev 71:829–830. https://doi.org/10.1103/PhysRev.71.829.5
7. CERN. 25 years of Large Hadron Collider experimental programme. https://home.cern/about/updates/2017/12/25-years-large-hadron-collider-experimental-programme. Accessed 07 Jan 2018
8. Django Manglunki (2013) Calcul, technique et réalisation des accélérateurs de particules. Course of 2nd master yead of civil engineering, ULB, given at CERN
9. Wikimedia Foundation. Linear accelerator for elementary particles with drift tubes. https://commons.wikimedia.org/wiki/File:Lineaer_accelerator_en.svg. Accessed 07 Jan 2018
10. CERN. LHC Machine Outreach. http://lhc-machine-outreach.web.cern.ch/lhc-machineoutreach/. Accessed 07 Jan 2018
11. Hanke K (2013) Past and present operation of the CERN PS Booster. Int J Mod Phys A28:1330019. https://doi.org/10.1142/S0217751X13300196
12. Pierre Freyermuth. LHC Report: imaginative injectors. https://cds.cern.ch/journal/CERNBulletin/2016/32/NewsArticles/2201549?ln=en. Accessed 08 Sept 2017
13. CMS glossary. https://twiki.cern.ch/twiki/bin/view/CMSPublic/WorkBookGlossary. Accessed 17 Apr 2017
14. Landshoff PV (2008) The Total cross-section at the LHC. Acta Phys Polon B39:2063–2094. arXiv:0709.0395 [hep-ph]
15. CMS Collaboration (2016) Measurement of the inelastic proton-proton cross section at $\sqrt{s} = 13$ TeV
16. CMS. CMS Public TWiki page for luminosity. https://twiki.cern.ch/twiki/bin/view/CMSPublic/LumiPublicResults. Accessed 30 Sept 2017
17. CMS Collaboration (2017) CMS luminosity measurements for the 2016 data taking period
18. CERN. CERN operation webtools. https://op-webtools.web.cern.ch. Accessed 08 Sept 2017

19. CMS Collaboration (2008) The CMS experiment at the CERN LHC. J Instrum 3(08):S08004. http://stacks.iop.org/1748-0221/3/i=08/a=S08004
20. Berardi V et al (2004) TOTEM: Technical design report. Total cross section, elastic scattering and diffraction dissociation at the Large Hadron Collider at CERN
21. Adriani O et al (2006) Technical design report of the LHCf experiment: measurement of photons and neutral pions in the very forward region of LHC
22. Pinfold J et al (2009) Technical design report of the MoEDAL experiment
23. Angelis ALS et al (2001) CASTOR: Centauro and strange object research in nucleus-nucleus collisions at the LHC. Nucl Phys B-Proc Suppl 97(1):227–230. ISSN 0920-5632. https://doi.org/10.1016/S0920-5632(01)01270-1, http://www.sciencedirect.com/science/article/pii/S0920563201012701
24. Armstrong WW, et al (1994) ATLAS: Technical proposal for a general-purpose p p experiment at the Large Hadron Collider at CERN
25. LHCb (1998) Technical Proposal. Technical Proposal. Geneva: CERN. https://cds.cern.ch/record/622031
26. ALICE (1995) Technical proposal for a large ion collider experiment at the CERN LHC
27. CMS, the Compact Muon Solenoid: Technical proposal (1994)
28. Tavernier S (2010) Experimental techniques in nuclear and particle physics. Springer, Berlin. https://doi.org/10.1007/978-3-642-00829-0
29. Davis SR (2016) Interactive slice of the CMS detector. https://cds.cern.ch/record/2205172
30. Chatrchyan S, et al (2014) Description and performance of track and primary-vertex reconstruction with the CMS tracker. JINST 9(10):P10009. https://doi.org/10.1088/1748-0221/9/10/P10009, arXiv:1405.6569 [physics.ins-det]
31. Emil Kálmán R (1960) A new approach to linear filtering and prediction problems. Trans ASME-J Basic Eng 82(Series D):35–45
32. Rose Kenneth (1998) Deterministic annealing for clustering, compression, classification, regression, and related optimization problems. Proc IEEE 86(11):2210–2239
33. Fruhwirth R, Waltenberger W, Vanlaer P (2007) Adaptive vertex fitting. J Phys G34:N343. https://doi.org/10.1088/0954-3899/34/12/N01
34. Chatrchyan S, et al (2013) Identification of b-quark jets with the CMS experiment. JINST 8:P04013. https://doi.org/10.1088/1748-0221/8/04/P04013, arXiv:1211.4462 [hep-ex]
35. Tanabashi M, et al (2018) Review of particle physics. Phys Rev D 98(3):030001
36. Bayatian GL, et al (2007) CMS physics: technical design report volume 2: physics performance. J Phys G 34 CERN-LHCC-2006-021. CMS-TDR-8-2. Revised version submitted on 2006-09-22 17:44:47, 995–1579, 669 p. http://cds.cern.ch/record/942733
37. Abdullin S, et al (2009) ERRATUM: The CMS barrel calorimeter response to particle beams from 2 to 350 GeV/c. 60:353–356
38. Cavallari F (2011) Performance of calorimeters at the LHC. J Phys Conf Ser 293(1):012001. http://stacks.iop.org/1742-6596/293/i=1/a=012001
39. Chatrchyan S, et al (2013) Energy calibration and resolution of the CMS electromagnetic calorimeter in pp collisions at $\sqrt{s} = 7$ TeV. JINST 8:[JINST8,9009(2013)], P09009. https://doi.org/10.1088/1748-0221/8/09/P09009, arXiv:1306.2016 [hep-ex]
40. Sirunyan AM, et al (2017) Particle-flow reconstruction and global event description with the CMS detector. JINST 12(10):P10003. https://doi.org/10.1088/1748-0221/12/10/P10003. arXiv:1706.04965 [physics.ins-det]
41. Khachatryan V, et al (2017) The CMS trigger system. JINST 12(01):P01020. https://doi.org/10.1088/1748-0221/12/01/P01020, arXiv:1609.02366 [physics.ins-det]

Chapter 4
Monte Carlo Techniques and Physics Generators

IN HEP, MONTE CARLO (MC) TECHNIQUES ARE USED in at least two different contexts: compute physics predictions, or simulate the interactions of particles with the detector. In this chapter, the basic techniques are presented. Then different physics generators, such as the ones used in Part II, are detailed.

4.1 Introduction

In this section, we want to illustrate the techniques used to produce predictions in HEP [1, 2]; in particular, we discuss:

– how to generate random numbers,
– and how to compute an integral such as the following:

$$I = \int_a^b f(u)\,\mathrm{d}u \qquad (4.1)$$

Indeed, the typical problem in HEP is to integrate a cross section such as in Eq. 2.2. The dimension of such a problem is $d = N_{\text{FS particles}} \times 3 - 4$, where 3 corresponds to the three coordinates of the momentum and 4 to the conservation of energy and momentum. At LHC, the final state is expected to contain a few hundreds of stable particles. In ultra-high-dimension problems, MC techniques are the only hope to achieve integration.

In this section, a few definitions and results in probability theory are recapitulated. Then the efficiency of Monte Carlo (MC) integration and sampling methods are discussed.

© Springer Nature Switzerland AG 2019
P. L. S. Connor, *Inclusive b Jet Production in Proton-Proton Collisions*, Springer Theses,
https://doi.org/10.1007/978-3-030-34383-5_4

4.1.1   Reminder on Probability Theory

Probability density function. The random variable X has probability density function (p.d.f.) g (non-negative, integrable and normalised to unity) if, for the probability $\mathbb{P}[a \leq X \leq b]$ of X being in the interval $[a, b]$, the following condition is satisfied:

$$\mathbb{P}[a \leq X \leq b] = \int_a^b g(x)\, dx \tag{4.2}$$

Expectation value and variance. Given f a function of a random variable X following a p.d.f. g, the *expectation value* $\mathbb{E}[f]$ and the *variance* $\mathbb{V}[f]$ are defined as follows:

$$\mathbb{E}[f] = \int_{-\infty}^{\infty} f(x)g(x)\, dx \tag{4.3}$$

$$\mathbb{V}[f] = \mathbb{E}\left[(f - \mathbb{E}[f])^2\right] \tag{4.4}$$

Law of Large Numbers (LLN). Given independent and identically distributed realisations x_i of the random variable X, for $N \to \infty$:

$$\frac{1}{N}\sum_{i=1}^{N} f(x_i) \longrightarrow \mathbb{E}[f] \tag{4.5}$$

In the particular case of a uniform distribution $u_i \sim \mathcal{U}[a, b]$, we have the simple form of the Law of Large Numbers (LLN) for $N \to \infty$:

$$\frac{1}{N}\sum_{i=1}^{N} f(u_i) \longrightarrow \frac{1}{b-a}\int_a^b f(u)\, du \tag{4.6}$$

4.1.2   Simple Monte Carlo Integration

Solution. A basic solution to compute Eq. 4.1 numerically is obviously given by the simple form of the LLN (Eq. 4.6):

$$\mathbb{E}[f] = \frac{1}{b-a}\int_a^b f(u)\, du \tag{4.7}$$

$$\approx \frac{1}{N}\sum_{i=1}^{N} f(u_i) \tag{4.8}$$

Hence we have an estimate for our integral:

$$I_{\text{MC}} = \frac{b-a}{N} \sum_{i=1}^{N} f(u_i) \qquad (4.9)$$

Using Eq. 4.4, the precision is given by the following:

$$\sigma_{\text{MC}}^2 = \mathbb{V}[I_{\text{MC}}] \qquad (4.10)$$

$$= \frac{(b-a)^2}{N} \mathbb{V}[f] \qquad (4.11)$$

$$= \frac{(b-a)^2}{N} \left(\frac{1}{N} \sum_{i=1}^{N} (f(u_i))^2 - \left(\frac{1}{N} \sum_{i=1}^{N} f(u_i) \right)^2 \right) \qquad (4.12)$$

Generation of uniformly distributed random numbers. The approach given in the previous paragraph assumes the existence of random generators. While such generators may be found in nature, based on random phenomena (abundant in quantum physics), they are intrinsically not reproducible. In numerical computation, one usually prefers pseudo-random generators, based on non-linear (but still deterministic) algorithms. Typically, a good generator is characterised by its capability to produce number very loosely correlated; but in general, assessing rigorously the quality of a generator is a non-trivial problem. A few examples of typical generators are illustrated in Fig. 4.1:

Congruential generator   Numbers are picked in a sequence of numbers u_i given
  by the following congruential relation:

$$x_{i+1} = ax_i + c \bmod m \qquad (4.13)$$

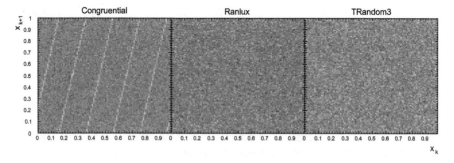

Fig. 4.1 Serial correlations in the generation of pseudo-random numbers with 10^7 entries. The congruential operator is given for $a = 205$, $c = 29573$, $m = 139968$ and $x_0 = 4711$. The pattern given by the congruential generator shows correlations, while the two other generators do not exhibit any

A suboptimal choice of the *multiplier a*, the *increment c* and the *modulus m* can lead to serial correlations.

RANLUX   Given the large amount of samples required in HEP, and given the increasing computing capabilities, specific random generators have been developed, such as RANLUX [3, 4]. This generator is based on the *Marsaglia–Zaman generator* [5] which can be considered as a congruential generator for a smart choice of the parameters (though with additional features), ensuring large periods of typically 10^{171}.

TRandom classes in ROOT   Other generators have also been developed, implemented among other in the ROOT library [6]. The *TRandom3* class implements an algorithm [7] remarkable for its particularly long period of $2^{19937} - 1$.

An application will be shown in Chap. 7.

4.1.3 Importance Sampling

A uniform sampling is not necessarily efficient when the function is peaked in a particular region of the domain. Therefore, one needs to modify the previous approach to improve the performance (Eq. 4.10 shows how the variance is related to the sample u_i).

Principle. Instead of the simple form, one introduces a p.d.f. g that mimics the function to integrate f, and uses the general form of the LLN (Eq. 4.5):

$$I = \int_a^b f(u)\, du \tag{4.14}$$

$$= \int_a^b \bar{f}(x)g(x)\, dx \tag{4.15}$$

$$= \mathbb{E}\left[\bar{f}\right] \tag{4.16}$$

Now, in order to estimate the integral with MC, one considers $\bar{f} = f/g$ as the integrand with the sample x_i distributed according to g:

$$I_{\text{MC}} = \frac{1}{N}\sum \bar{f}(x_i) \tag{4.17}$$

Generation of non-uniformly distributed random numbers. Given $u_i \sim \mathcal{U}[0, 1]$, the sample x_i can be described by the p.d.f. g if the following applies:

$$\int_{-\infty}^{x_i} g(x)\, dx = u_i \int_{-\infty}^{+\infty} g(x)\, dx \tag{4.18}$$

Intuitively, for u_i uniformly distributed, x_i will be mostly located where g is sharp. As an application, in the PS (see Sect. 2.3.3.2), the next branching is determined by generating a random number according to a Sudakov factor for each outgoing leg that has not reached the hadronisation scale yet: the random number corresponding to the leg with the highest scale is chosen to describe the next branching, and the other random numbers are discarded; the procedure is then iterated, as described previously.

4.2 Physics Generators

MC techniques are widely used in HEP to produce theoretical predictions. From the large variety of different physics generators, we review some of them here.

4.2.1 Fixed-Order Parton-Level Calculations

Parton-level calculations are based on the factorisation (Eq. 2.37), and only include the PDFs the ME at fixed order (FO) in the perturbative regime. Typically, LO and NLO can be achieved, though some NNLO calculations can sometimes be found for certain processes. However, their use is somehow limited:

– they do not produce full event records similar to the data;
– missing orders can lead to sizeable discrepancies in the predictions for small-cone size jet clustering algorithms;
– finally, they do not include hadronisation.

At the time of writing this thesis, standard references of such generators are the following:

MADGRAPH   MADGRAPH(or MADEVENT — both names shall here be used as synonymous) is a generator at LO and at NLO, including however more diagrams at tree level with an arbitrary number of particles in the final state [8–10]. In this thesis, we use MadGraph5_aMC@NLO 2.2.2 at LO for $2 \to 2$, $2 \to 3$ and $2 \to 4$ processes.

POWHEGbox   POWHEG is the method described in Refs. [11–13], and the name stands for *POsitive WeigHts Event Generator*. In this thesis, we use it for the production of QCD dijet processes with $2 \to 2$ and $2 \to 3$ at NLO. The additional radiation is treated with Sudakov factor using a splitting function at NLO. The POWHEG box corresponds to an implementation of the POWHEG method [14]. Usually, it is not used standalone but is interfaced with an event generator, as described in the next section.

NLOJET++   NLOJET++ is a "pure" fixed-order calculation, in the sense that it is only implement Eq. 2.37. No parton shower can be applied and the result is at parton-level.

4.2.2 Event Generators

MC event generators simulate particle collisions on an event-by-event basis, similarly to what happens in the real detector. Generated events include the Matrix Element (ME) and the Underlying Event (UE). The latter can further decomposed as follows:

– extra ISRs and FSRs,[1]
– Multi-Parton Interaction (MPI),
– hadronisation,
– hadron decays and soft photon radiations,
– and treatment of beam remnants.

This reflects the current understanding of an interaction, described in Sect. 2.3. The main difficulty consists in connecting the different steps in a consistent way, i.e. matching the ME and the Underlying Event (UE).

4.2.2.1   General-Purpose Event Generators

Today, three General-Purpose MC Event Generator (GPMC) make reference, including the simulation of different types of scatterings (proton-proton, electron-positron, electron-proton, etc.), different types of interaction (EW, QCD, Higgs processes, processes BSM, etc.):

PYTHIA   Several versions of PYTHIA exist, the most recent being PYTHIA 8, written in C++ [15, 16], which is used in this thesis[2] (version 8.205). The hard process is hard-coded at the Born level, i.e. at LO. The generator includes p_T-ordered PS with *angular veto* and Lund string hadronisation. The MPI is accounted for with a smoothing factor with parameter p_{T0}. It is also interesting to note that PYTHIA makes use of the Marsaglia–Zaman algorithm presented above for the generation of random numbers.

HERWIG   For this generator as well, two versions are maintained in parallel[3]: HERWIG++ and HERWIG 7, both written in C++ [17]. HERWIG 7 implements automated NLO calculations, while in HERWIG++, the hard process is hard-coded at the Born level. Unlike PYTHIA, HERWIG uses *angular-ordering* for PS and *cluster model* for hadronisation. The MPI is accounted for with a sharp cut-off with parameter p_{T0}. In this thesis, only HERWIG++ is considered.

SHERPA   SHERPA includes automated calculation of MEs at LO and at NLO [18]. It includes its own PS and UE. Similarly to PYTHIA, the MPI is performed with a smoothing factor. It is here only mentioned and not further discussed in this thesis.

---

[1] Depending on the context, the PS is sometimes considered separately from the UE.

[2] One should mention PYTHIA 6, previous version written in FORTRAN, no longer actively maintained. However, PYTHIA 8 is largely based on PYTHIA 6, whose manual stays an important reference.

[3] Here, one should also mention the previous version HERWIG 6, written in FORTRAN, but which is also no longer maintained.

General-Purpose MC Event Generator (GPMC) can also be interfaced with some of the fixed-order calculations [19], like MADGRAPHand POWHEG, as will be the case in Part II of this thesis, or with other event generators that only take care of a part of the full chain.

One interesting example is CASCADE, an event generator using off-shell MEs with CCFM evolution [20]; it includes its own ISR but needs to be interfaced with, for instance, PYTHIA for the FSR and for hadronisation.

4.2.2.2 Tuning

The phenomenological models used in the UE involved parameters, which need to be estimated, or *tuned*. A non-exhaustive list of the parameters is given in Table 4.1. A *tune* corresponds to a set of parameters of the UE.

In order to determine the parameters, a fit of to data of several MB distributions (described later in this section) can be performed for different values of the parameters. A χ^2 is then computed and the parameters providing the best agreement is selected:

$$\chi^2(\mathbf{p}) = \sum_{\text{observables}} w_{\text{observable}} \sum_{\text{bins}} \left(\frac{f_{\text{bin}}(\mathbf{p}) - \mathcal{R}_{\text{bin}}}{\Delta^2_{\text{bin}}} \right)^2 \tag{4.19}$$

where

- $\mathbf{p}$ stands for the free parameters;
- $\mathcal{R}$ stands for the data (*reference*);
- f stands for the model (hadronisation, PS, etc.);
- Δ_{bin} stands for the experimental uncertainties;
- and w stands for the weight.

The weight is chosen more or less arbitrarily in order to increase the impact of a certain distribution with respect to others.

Table 4.1 Non-exhaustive list of parameters in phenomenological models that need to be tuned (exhaustive list given in Chap. 9)

Model	Parameter	Signification
PS	$\alpha_S(M_Z)$	Strong coupling
	p_{T0}	Smoothing parameter
MPI	R	Involved in *colour reconnection*
	p_{T0}	Smoothing parameter
Hadronisation	a, b	Flavour-dependent, only for Lund string fragmentation
	r_Q	Bowler parameter, for HF
	Q_0	Hadronisation scale
Intrinsic k_T	σ	Width of the Gaussian distribution

Fig. 4.2 Definitions of the different tuning regions in tuning techniques: the *toward* (*away*) region correspond to the direction (opposite direction) of the leading object; the tuning is performed in the transverse regions. The *trans max* and *trans min* are defined by the presence of a third jet, to disentangle further the contributions to the UE

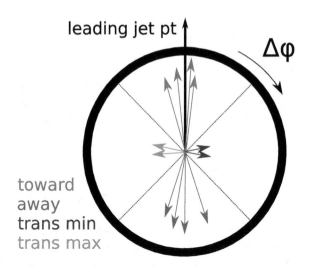

The naïve approach consists in performing a grid scan of all possible parameters; however, this method is extremely time- and resource-consuming. In the modern approach, the simulation is run in parallel for randomised sets of parameter [21]; the value of the parameters is then determined by interpolations. The current reference for this method in HEP is called PROFESSOR (PROcedure For EStimating Systematic errORs) [22, 23].

The activity coming from the UE—the *soft* activity—is disentangled from hard interaction by selecting the activity in the region transverse to the direction of the leading-p_T particle, as illustrated in Fig. 4.2. The considered observables are the multiplicity of the charged particles and the scalar sum of the transverse momenta of the charged particles. Sometimes, in order to disentangle contributions from the PS and from the UE, the transverse region is further separated into the *transMAX* region with the presence of a third jet and the *transMIN*.

In Part II, two tunes will be used [10]:

– CUETP8M1 (CMS UE Tune PYTHIA 8 Monash 1)
– CUETHppS1 (CMS UE Tune HERWIG++ Set 1)

They only differ in the physics generator that is used, but follow the same treatment:

– They are both based on the tunes obtained from previous experiments. For instance, the hadronisation parameters are still corresponding to parameters obtained at the LEP in $e^+e^- \to Z \to q\bar{q}$.
– For parameters related to hadronic collisions (like MPI or intrinsic k_T), MB data samples from Tevatron at $\sqrt{s} = 900$ Gev and 1.96 Tev and from LHC at $\sqrt{s} = 7$ Tev are used.

The interest of b jet measurements will be discussed in Chap. 5.

4.3 Detector Simulation

One major difference still remains between the event records in physics generators and in data: the former contains the exact kinematics of the full list of stable and unstable particles, while the latter contains lists of tracks and of energy deposits. In order to compare data and simulation, one needs either to apply the effect of the detector (i.e. simulate the interactions of the particles with the different subdetectors) or to correct for it in the data. In both cases, the simulation of the detector is needed.

One software among all makes reference for the simulation of interactions of high-energy particles with media: GEometry ANd Tracking (GEANT4) [24–26]. It is used for the simulation of most of the modern experiments in HEP such as CMS, as well as in many experiments in space science, medical science, and engineering.

Given the description of the detector, which encompasses not only the active but also the passive parts of the detector, GEANT4 simulates the trajectories of all outgoing particles. In the description of the interaction of particles with the detector, random number generators are typically of use in the simulation of *multiple scattering* or of *decays*.

Although the simulation of the detector is given with very good precision, it is not perfectly correct. The *response* of the detector will present differences between the data and the simulation; for instance, in practice, the time dependence is not simulated.

Conclusion

MC techniques are heavily used in HEP at all levels. In Part II, MC simulations will be used in different contexts:

- First, the effect of the detector will be studied thanks to simulations based on MC samples (Chap. 7).
- Then, the effect of the detector on data will be removed thanks to the simulations (Chap. 8).
- Finally, the corrected data will be compared to predictions obtained from MC samples (Chap. 9).

However, it is important to stress that simulations only offer an approximative description of nature. Simulations are to be handled with care when comparing to measurements and when utilising them to correct the measurement from detector effects. An important part of the difficulty of a physics analysis relies in the assessment the description of the measurement by the simulation.

4.A Further Methods of MC Integration

Additional methods exist to compute efficiently integrals, by subdividing the region to integrate.

Stratified sampling. Here, the region is divided horizontally by considering successive intervals:

$$\int_a^b f(x)\,dx = \int_a^c f(x)\,dx + \int_c^b f(x)\,dx \qquad (4.20)$$

Each term can then be integrated with a different p.d.f. In Part II, several examples will be encountered with the samples used to study the transverse momentum, which is a steeply falling spectrum, in order to have a large enough statistical sample also at high values:

- The PYTHIA-8 sample is produced in slices of the transverse momentum $\hat{p}_T$ of the outgoing partons, as illustrated in Fig. 4.3.
- Similarly, the MADGRAPH sample is produced in slices of scalar sum of the transverse momenta of the hard partons in the final state $H_T = \sum_i p_{Ti}$.

Subtraction method. Here, the region is divided vertically on the same interval by considering different p.d.f. f and g:

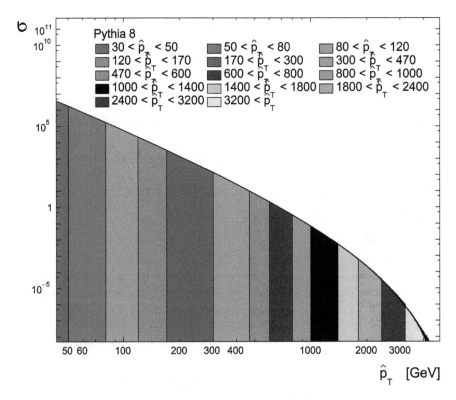

Fig. 4.3 The PYTHIA-8 samples are generated in 14 bins of $\hat{p}_T$, corresponding to the transverse momentum of the hard partons in the final state. A similar statistical precision can therefore be achieved over several orders of magnitude

$$\int_a^b f(x)\,dx = \int_a^b g(x)\,dx + \int_a^b (f(x) - g(x))\,dx \qquad (4.21)$$

This method is typically used in fixed-order calculations. For instance, while in MC@NLO, some events have negative weight, terms are rearranged in POWHEG such that all entries are positive (hence its name).

Hit-or-Miss integration. This integration method is an alternative to Eq. 4.17. The principle is to pick N couples of random numbers x_i, y_i in a rectangle:

$$a \leq x_i \leq b \qquad (4.22)$$
$$0 \leq y_i \leq c \qquad (4.23)$$

Determine the number M of couples such that $y_i \leq f(x_i)$. The integral is then estimated as follows:

$$I \approx c(b - a)\frac{M}{N} \qquad (4.24)$$

A example of application of this algorithm can be found in the treatment of the PS, while dealing with the Sudakov factor.

References

1. James F (1980) Monte Carlo theory and practice. Repo Prog Phys 43(9):1145. http://stacks.iop.org/0034-4885/43/i=9/a=002
2. Cowan G (1998) Statistical data analysis. Oxford University Press, Oxford
3. Lüscher M (2016) RANLUX. http://luscher.web.cern.ch/luscher/ranlux/guide.pdf
4. Luscher M (1994) A portable high quality random number generator for lattice field theory simulations. Comput Phys Commun 79:100–110. https://doi.org/10.1016/0010-4655(94)90232-1, arXiv:hep-lat/9309020 [hep-lat]
5. Marsaglia G, Zaman A (1991) A new class of random number generators. Ann Appl Probab 1(3):462–480. https://doi.org/10.1214/aoap/1177005878
6. Brun R, Rademakers F (1997) ROOT–An object oriented data analysis framework. Nucl Instrum Methods Phys Res Sect A: Accel Spectrom Detect Assoc Equip 389(1). New Computing Techniques in Physics Research V, pp 81–86. ISSN: 0168-9002. https://doi.org/10.1016/S0168-9002(97)00048-X
7. Matsumoto M, Nishimura T (1998) Mersenne twister: a 623-dimensionally equidistributed uniform pseudo-random number generator. ACM Trans Model Comput Simul 8(1):3–30. ISSN: 1049–3301. https://doi.org/10.1145/272991.272995
8. Alwall J et al (2011) MadGraph 5: going beyond. JHEP 06:128. https://doi.org/10.1007/JHEP06(2011)128, arXiv:1106.0522 [hep-ph]
9. Alwall J et al (2014) The automated computation of tree-level and next-to-leading order differential cross sections, and their matching to parton shower simulations. JHEP 07:079. https://doi.org/10.1007/JHEP07(2014)079, arXiv:1405.0301 [hep-ph]
10. Khachatryan V et al (2016) Event generator tunes obtained from underlying event and multiparton scattering measurements. Eur Phys J C76(3):155. https://doi.org/10.1140/epjc/s10052-016-3988-x, arXiv:1512.00815 [hep-ex]

11. Alioli S et al (2011) Jet pair production in POWHEG. JHEP 1104:081. https://doi.org/10.1007/JHEP04(2011)081, arXiv:1012.3380 [hep-ph]

12. Alioli S et al (2010) A general framework for implementing NLO calculations in shower Monte Carlo programs: the POWHEG BOX. JHEP 06:043. https://doi.org/10.1007/JHEP06(2010)043, arXiv:1002.2581 [hep-ph]

13. Nason PA (2010) Recent developments in POWHEG. In: Proceedings, 9th international symposium on radiative corrections: applications of quantum field theory to phenomenology (RADCOR 2009), vol RADCOR2009, p 018. arXiv:1001.2747 [hep-ph]

14. Ježo T, Nason P (2015) On the treatment of resonances in next-to-leading order calculations matched to a parton shower. JHEP 12:065

15. Sjöstrand T, Mrenna S, Skands P (2006) PYTHIA 6.4 physics and manual. JHEP 05:026. https://doi.org/10.1088/1126-6708/2006/05/026, arXiv:hep-ph/0603175 [hep-ph]

16. Sjöstrand T, Mrenna S, Skands P (2008) A brief introduction to PYTHIA 8.1. Comput Phys Commun 178(11):852–867. ISSN: 0010-4655. https://doi.org/10.1016/j.cpc.2008.01.036, http://www.sciencedirect.com/science/article/pii/S0010465508000441

17. Bahr M et al (2008) Herwig++ physics and manual. Eur Phys J C 58:639. https://doi.org/10.1140/epjc/s10052-008-0798-9, arXiv:0803.0883 [hep-ph]

18. Gleisberg T et al (2009) Event generation with SHERPA 1.1. J High Energy Phys 2009(02):007. http://stacks.iop.org/1126-6708/2009/i=02/a=007

19. Boos E et al (2001) Generic user process interface for event generators. In: Physics at TeV colliders. Proceedings, Euro summer school, Les Houches, France, May 21–June 1, 2001. http://lss.fnal.gov/archive/preprint/fermilab-conf-01-496-t.shtml, arXiv:hepph/0109068 [hep-ph]

20. Jung H (2002) The CCFM Monte Carlo generator Cascade. Comput Phys Commun 143(1):100–111. ISSN: 0010-4655. https://doi.org/10.1016/S0010-4655(01)00438-6

21. Abreu P et al (1996) Tuning and test of fragmentation models based on identified particles and precision event shape data. Z Phys C73:11–60. https://doi.org/10.1007/s002880050295

22. Buckley A et al (2010) Systematic event generator tuning for the LHC. Eur Phys J C65:331–357. https://doi.org/10.1140/epjc/s10052-009-1196-7, arXiv:0907.2973 [hep-ph]

23. Buckley A et al (2018) Tutorial on PROFESSOR. https://professor.hepforge.org/prof-tutorial.pdf. Accessed: 2018-01-01

24. Agostinelli S et al (2003) Geant4-a simulation toolkit. Nucl Instrum Methods Phys Res Sect A: Accel Spectrom Detect Assoc Equip 506(3):250–303. ISSN: 0168-9002. https://doi.org/10.1016/S0168-9002(03)01368-8

25. Allison J et al (2006) Geant4 developments and applications. IEEE Trans Nucl Sci 53:270. https://doi.org/10.1109/TNS.2006.869826

26. Allison J et al (2016) Recent developments in Geant4. Nucl Instrum Methods Phys Res Sect A: Accel Spectrom Detect Assoc Equip 835(Supplement C):186–225. ISSN: 0168-9002. https://doi.org/10.1016/j.nima.2016.06.125, http://www.sciencedirect.com/science/article/pii/S0168900216306957

Chapter 5
Discovery, Overview and Motivation of Beauty Physics

Why is bottom quark physics so interesting? The cynic might argue that the labs are into bottom quark physics because it's affordable.
—Edward H. THORNDIKE [1]

IN THE FIRST PART OF THIS CHAPTER, the discovery and current knowledge of the properties of the *b* quark are reviewed. Then the current status and prospects for deepening this knowledge with the LHC experiments are detailed. Finally, tagging techniques and measurements of *b* quarks as a probe are discussed.

5.1 Discovery and First Measurements

5.1.1 The *E288* Experiment at Fermilab

In the seventies, several experiments at Fermilab were ongoing to search for new resonances, especially the $W^\pm$ and Z^0 bosons of the electroweak theory. In this context, the *bottom* quark—or more poetically *beauty* quark—was discovered in 1977 with a fixed-target experiment at Fermilab led by Leon LEDERMAN in *bottomonium* states $b\bar{b}$ [2–5].

In the measurement of the invariant mass of dimuon systems from the outgoing particles of the collisions of 400-GeV protons on a nuclear target, they observed a double-peak structure around 9.5 GeV (see Fig. 5.1). The situation being very analogous to the discovery of J/ψ [7, 8], i.e. a very narrow resonance, the new discovery

© Springer Nature Switzerland AG 2019
P. L. S. Connor, *Inclusive b Jet Production in Proton-Proton Collisions*, Springer Theses,
https://doi.org/10.1007/978-3-030-34383-5_5

Fig. 5.1 The plot showing
an excess of events around
9.50 GeV, marking the
discovery of the Υ. « There
was no known object that
could explain that bump,
»Leon LEDERMAN said,
E288 spokesman and
Fermilab director
emeritus [6]

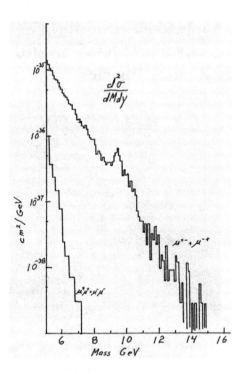

was similarly interpreted as bound systems of quark-antiquark pairs of a new generation, this time with charge $-1/3$, called Υ's.[1]

The process of production and decay of $b\bar{b}$ states can be described as follows:

$$p + N \rightarrow \Upsilon \rightarrow b\bar{b} \rightarrow \gamma^* \rightarrow \mu^+\mu^- \tag{5.1}$$

where N stands for copper, platinum or beryllium; this corresponds to an electromagnetic decay. While the existence of a new quark seemed clear, its properties, such
as electroweak properties, could not be further determined; therefore this production
mode of b quarks is called *hidden beauty* production. Confirming its existence and
determining further its properties in the production of *open beauty* motivated further
investigations on this new quark in later experiments.

But in addition, in contrast to the discovery in 1974 of the *charmonium* J/ψ that
was expected (with the *GIM mechanism* [9, 10]) and understood as a pair of heavy
quarks [11], the discovery in 1978 of the bottomonium Υ was rather a surprise, since

---

[1] Two legends exist regarding the name of the particle. The first is that "Upsilon" stand for "up+psi",
in reference to the J/ψ. The other legend is related to the spokesperson and leader of team's name:
the Nobel-prized Leon LEDERMAN. Observing first a resonance at 6 GeV in 1976, the members of
the team agreed that its name would be Upsilon if it would be established, or *Oops-Leon* if it would
not. The resonance at 6 GeV turned out to be nothing else than a fluctuation; however, using the
same experimental set-up, they found a new resonance at 9 GeV: this time a real resonance. They
decided to keep the name Upsilon.

a third generation of quarks was not yet evident. So far, only the famous publication on CP violation by Makoto KOBAYASHI and Toshihide MASKAWA [12] was assuming six quarks. The *tauon* was also not well established, which would have led to more serious speculations on a third generation of quarks. In addition, its discovery raised new questions, such as the existence of a partner to the fifth quark.

5.1.2 ISR

Situated at CERN, the Intersecting Storage Rings (ISR) was the first hadron collider in the world, composed of two rings with 150 m of diameter. It could deliver proton-proton and $p\bar{p}$ collisions at a centre-of-mass energy up to 62 GeV. It was active from 1971 to 1984 [13].

At the end of 1977, ISR confirmed the discovery of the Υ with proton-proton collisions in the di-electron invariant-mass spectrum [14]:

$$pp \to \Upsilon X \to e^+ e^- X \tag{5.2}$$

The mass was estimated to $m_\Upsilon = 9.46 \pm 0.16\,\text{GeV}$, in agreement with the findings of the E288 experiment.

5.1.3 DORIS

DORIS[2] was one of the first *storage rings*, allowing e^+e^- collisions, with a circumference of 289 m. Built from 1969 to 1974 at *Deutsches Electroknen-Synchrotron* (DESY), each beam was carrying 3.5 GeV of energy in the lab frame. If this was enough to study the production of charmonium, this was too low to reproduce the new resonance discovered at Fermilab. After undergoing an upgrade in 1978, beams could reach an energy of 5 GeV, which was then enough to confirm the findings of Fermilab E288 experiment.

Three experiments at DORIS confirmed the resonance in the same year: PLUTO, DASP2 and DHHM.

The PLUTO collaboration[3] confirmed the existence of the resonance in 1978 [15] through its electromagnetic decay (Eq. 5.1) in hadronic production: whereas the decay into a dimuon system accounts for around 3% of the decays, hadronic decays account for 90%, and the most significant channel is the production of three hadrons:

$$e^+ + e^- \to \Upsilon \to b\bar{b} \to ggg \to hhh \tag{5.3}$$

---

[2]*DOppel-RIng Speicher.*
[3]Named after the magnet of the detector.

The PLUTO experiment measured $m_\Upsilon = 9.46 \pm 0.01$ GeV and confirmed from the measurement that, in analogy with J/ψ, it should be made of a pair of quarks of charge $-1/3$ [11].

In addition, PLUTO confirmed the discovery of the tauon, first observed at the SPEAR[4] in the years 1974–1977. Indeed, the existence of a third generation of leptons seriously supports the existence of a third generation of quarks.

The Double Arm SPectrometer (DASP2) experiment published simultaneously as PLUTO [16, 17]. The measurement was simpler, since it was only relying on the direction of the particles. In addition, the measurement was also performed through the electromagnetic decay:

$$e^+ + e^- \to \Upsilon \to \mu^+\mu^- \tag{5.4}$$

A few months later, the DESY-Hamburg-Heidelberg-München collaboration (DHHM) experiment also announced it, together with an excited state at 10.02 ± 0.02 GeV [18], similarly to E288, resolving then the double-peak structure previously observed.

The *Cornell potential*, already used to describe the $c\bar{c}$ system [19], is also tested to describe the $b\bar{b}$ system. This potential is composed of two contributions:

$$U(r) = \underbrace{-\frac{a}{r}}_{\text{Coulombic part}} + \underbrace{br}_{\text{confinement}} \tag{5.5}$$

where

- the Coulombic part describes the one-gluon interaction in analogy to the Coulomb interaction;
- the confinement part includes the non-perturbative, not well understood effects.

This empirical potential[5] is successfully used for charm and bottom.

5.1.4 CESR

After the successes at E288 and at DORIS, there could be no more doubt about the existence of the fifth quark. Yet it was still observed in its hidden form, i.e. in the form of Υ mesons. Another topic of interest was the spectroscopy: how similar to J/ψ was Υ? In particular, should one expect to find an excited state decaying in B mesons, analogously to D mesons? Finally, what about a sixth quark?

At the *Cornell's laboratory for Nuclear Physics*, a new, 768-m long, symmetric e^+e^- collider was being built at the time of the discovery, the Cornell

---

[4] *Stanford Positron Electron Asymmetric Rings*, situated at the SLAC.

[5] One obvious limitation of this potential is that it does not allow the *fragmentation* of a pair of quarks. However, it successfully allowed to perform spectroscopy and describe lifetime.

Table 5.1 B mesons' content, mass and lifetime. The $B^{\pm}$ and B^0 are from far the most common ones, and have almost the same mass and very similar lifetimes. B^+ and B^- are one another's respective antiparticles. Values are taken from PDG [24]

Meson	Quark content	$M/$ MeV	$\tau/$ps
$B^{\pm}$	$u\bar{b}, \bar{u}b$	5279.29 ± 0.15	1.638 ± 0.004
B^0	$d\bar{b}$	5279.61 ± 0.16	1.520 ± 0.004
$B_c^{\pm}$	$c\bar{b}, \bar{u}b$	6275.1 ± 1.0	0.507 ± 0.009
B_s^0	$s\bar{b}$	5366.79 ± 0.23	1.510 ± 0.005

Electron-positron Storage Ring, pronounced "Caesar" (CESR), with two detectors: *Cleopatra*, for her/its, proximity with Caesar (CLEO) and Columbia University-Stony Brook (CUSB). Their contributions to HF physics would be important: a general review of Υ physics at CESR may be found in [20]; here, only a few key steps are mentioned.

Just a few months after the commissioning of the experimental set-up, the three first resonances were found before end of 1979 [21].

But more interestingly, in Spring 1980, a fourth resonance was found around 10.5 GeV by both experiments [22, 23] (Fig. 5.2). The peak was much broader, indicating a faster decay; this was understood, in analogy to the decay of J/ψ in pairs of D mesons, as a decay of Υ in a new kind of mesons, named B mesons:

$$e^+e^- \rightarrow \Upsilon(4S) \rightarrow B + \bar{B} \quad \text{(strong int.)} \tag{5.6}$$

where B is a B meson. The characteristics of B mesons are summarised in Table 5.1: their content, mass and lifetime are given.

Then the B meson will further decay weakly:

$$B \rightarrow XW \rightarrow Xl\nu \quad \text{(weak int.)} \tag{5.7}$$

A peak being found in the outgoing-lepton spectrum at this energy, "bare bottom" was found. The door to a new area of physics was definitely open.

5.2 Further Investigations

Since its discovery, the Υ and B mesons have kept physicists busy for several reasons:

– First, the electroweak properties related to the existence of the bare bottom, typically the CKM matrix and the CP violation, are heavily studied at Cornell Electron–positron Storage Ring, pronounced "Caesar" (CESR) and in following experiments (Positron-Electron Project (PEP), DORIS-II, . . .).

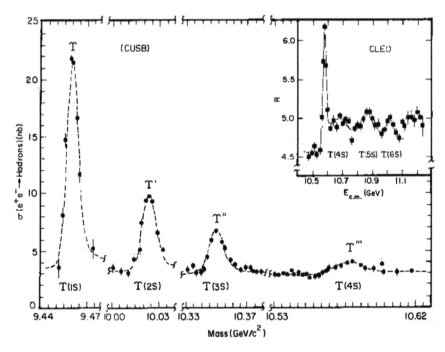

Fig. 5.2 Four Υ peaks at both CLEO and CUSB [25]. The three first peaks may be explained in terms of $b\bar{b} \to ggg$, $b\bar{b} \to \gamma gg$ or $b\bar{b} \to \gamma* \to q\bar{q}/l^+l^-$. In the fourth peak, one should also consider $b\bar{b} \to B\bar{B}$ where $B = b\bar{q}$ ($\bar{B} = \bar{b}q$)

- The partner of the beauty quark, the *top*—or *truth*[6]—is found only in 1995 at TEVATRON by the DØ[7] and Collider Detector at Fermilab (CDF)[8] collaborations, confirming many predictions obtained from measurements of b and B properties.
- Moreover, the study of the quarkonium $b\bar{b}$ (together with $c\bar{c}$) continued, giving rise to the Υ *spectroscopy*.
- B mesons are investigated, in particular for their remarkably long lifetimes. Also, other mesons, B_s and B_c, combining b with s or c quarks, are observed.

Since most of the modern particle accelerators have allowed the study of b quarks and B mesons, an exhaustive review is beyond the scope of this thesis; an overview of the discoveries in b physics to estimate size up its importance is given in Table 5.2, summarising the most notable contributions, experiments and colliders. Among

---

[6]This appellation has hardly gone beyond the joke "there is no truth in the SM" while physicists were struggling at finding it.

[7]Named after the location of the detector.

[8]Named Collider Detector at Fermilab.

Table 5.2 Main experiments contributing to b physics are summarised. The experiments are ordered by collider and starting of activity (many experiments were contemporary). A non-exhaustive list of the contributions of each experiment is given

Lab	Collider	Beams	$\sqrt{s}$	Experiment	Activity	Contributions	Further references
Fermilab		pN	≤10 GeV	E288		Discovery	[4]
CERN	ISR	pp	≤62 GeV	(Only one)	1971–1984	Confirmation	[13, 14]
DESY	DORIS-I	e^+e^-	~10 GeV	PLUTO		Confirmation	[15]
				DASP2		Confirmation	[16, 17]
				DHHM		Third resonance	[18]
Cornell	CESR	e^+e^-	~10 GeV	CLEO	1979–2003	Υ and B spectroscopy and measurement of $\|V_{cb}\|$	[27]
				CUSB	1979–1989	*Idem*	[28]
SLAC	PEP-I		≤29 GeV	(Several)	1980–1990	First measurements of B lifetime	[29]
DESY	DORIS-II	e^+e^-	~10 GeV	ARGUS	1982–1992	$B^0 - \bar{B}^0$ mixing. Suspicions toward very massive top quark	[30–32]
CERN	Sp$\bar{\text{p}}$S	$p\bar{p}$	546–630 GeV	UA1	1981–1990	$B^0 - \bar{B}^0$ mixing. Searches	[33–36]
Fermilab	TEVATRON	$p\bar{p}$	1.8 TeV	DØ	1983–2011	Production modes. Top quark	
				CDF	1983–2011	Production modes. Top quark $B_s - \bar{B}_s$ mixing. Secondary-vertex tagging	[37–39]
				B-TeV	(Aborted)		
SLAC	SLC	e^+e^-	~90 GeV	Mark-II	1982–1985	b fragmentation	
				SLD	1992–1998	b fragmentation	
	PEP-II	e^+e^-	~10 GeV	BaBar	1999–2008	CP violation. Secondary-vertex tagging	
DESY	HERA	ep	318 GeV	H1	1992–2007	b mass and running	[40]
				ZEUS	1992–2007	b mass and running	[40, 41]
				HERA-B	2002–2003	Novel method of production of B mesons. CP-violation	
CERN	LEP	e^+e^-	45–209 GeV	(Several)	1989–2000	B lifetime, b fragmentation	[42–44]
KEK	KEKB	e^+e^-	~10 GeV	Belle	1999–2010	B physics. CP violation	[45, 46]
CERN	LHC	pp	7, 8, 13 TeV	CMS	2009–today	B physics. CP violation. Searches	[47–52]
				ATLAS	2009–today	*Idem*	[47, 53–55]
				LHCb	2009–today	B physics. CP violation	[47, 56–59]
		Pb-Pb		ALICE	2009–today	B-physics in ion collisions	[60]

all experiments, some of them were especially dedicated to the study of Υs and *B* mesons, called *B factories*,[9] operating exactly at the mass of $\Upsilon(4S)$: PEP-II, Belle [26][10] and HERA-B.[11]

5.3 Physics at Hadron Colliders with *b*'s

At LHC, following up on the physics research at TEVATRON as a hadron collider, properties of *b* quarks are further investigated in order to refine previous results and validate the current understanding of QCD.

Given its remarkable, long lifetime, *b* quarks or *B* mesons are in principle easy to detect in *pp* collisions. Assuming their properties to be known, they have become an interesting probe for other processes in and beyond the SM.

This section is organised in four subsections: in the first subsection, we define the flavour of jets and investigate tagging techniques; then the mechanism of production of *b* jets is detailed; finally, in the two last subsections, applications for some important processes where *b* quarks play a rôle are outlined. Throughout this section, we try to highlight relevant aspects for the analysis presented in Part II.

5.3.1 Heavy-Flavour Jets

5.3.1.1 Definition

At hadron colliders such as LHC, *b* quarks are mostly collimated inside of jets. The *flavour* of a jet can be defined as follows:

1. according to the *hadron-flavour definition*, *b* jets must contain a stable *B* meson;
2. similarly, *c* jets must contain a stable *D* meson, unless already accounted as *b* jets;
3. and all other jets are defined as light jets (or *udsg*).

(We do not discuss here the possibility to define *t* jets.) Whether a particle is considered as stable is sometimes matter of debate and the definition at generator level can change:

- At TEVATRON and during LHC Run-I, the standard flavour definition was performed at parton level, with *b* and *c* quarks.
- During LHC Run-II (as for the present work), the definition was at hadron level, and *stable particles* were defined with respect to all strong and electroweak interactions.

---

[9]One should maybe also mention an attempt of *B* factory that failed, B- TEV, because of termination by the U.S. government.

[10]French word for beauty.

[11]HERA Beauty.

– In other experiments (e.g. ZEUS[12] and H1[13]), the flavour is also defined at hadron level, with the difference that the stability is only required for strong and electromagnetic interactions but *not* for weak interactions [61, 62].

Despite these remarks, it is common to keep the notation *c* jet or *b* jet (i.e. with small letters) even with the hadron-flavour definition; since this is the convention at CMS, we shall follow it here.

5.3.1.2 Tagging

The object of tagging is to identify, or *tag*, *b* jets, i.e. how to discriminate *b* jets from *c* and light jets. Different types of discriminants exist, based on different properties of *b* jets.

Indeed, the intermediate lifetime and masses of *B* hadrons lead to distinctive kinematic signatures in the tracks produced within *b* jets. In addition, HF jets have an increased probability to contains leptons.

LHC experiments rely on similar approaches [63–66], often combined with Multi-Variate Analysis (MVA).

A sketch of an event containing a *b* jet is shown in Fig. 5.3; the three tagging techniques are illustrated:

secondary vertex   The long lifetime of the *B* mesons imply its decay to be significantly displaced with respect to the PV (but usually still inside of the beam pipe). For instance, at CMS, the mean free path for heavily boosted *B*'s is of the order of $\lambda \approx 2$ mm. Most modern taggers rely primarily on the presence of a SV.

impact parameter   As a consequence of the presence of a SV, the impact parameter of the tracks with respect to the PV should be significantly larger. A jet quantity based on the impact parameter of its tracks should therefore be sensitive to the presence of a HF quark or hadron.

soft leptons   The weak decay of a *B* meson may lead to a non-isolated lepton in the final state:

$$B \rightarrow WX \rightarrow Xl\nu \tag{5.8}$$

In practice, in the final state of *pp* collisions at LHC, Υ are much less likely to take place, but they could also lead to non-isolated leptons in the final state:

$$\Upsilon \rightarrow ll \tag{5.9}$$

---

[12] Named in reference to the relation of Zeus and Hera in the mythology.
[13] HERA-1.

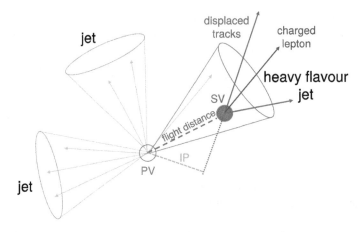

Fig. 5.3 Sketch of an event with b jets in pp collisions: the Primary Vertex (PV) is represented at the centre of the sketch with a small circle, while the Secondary Vertex (SV) is represented inside of the HF jet with a plain disc; jets are represented with cones pointing to the centre of the sketch, the HF jet with thicker lines; the tracks are represented with arrows. The three characteristic allowing the distinction of the jets with respect of the flavour are the interaction point (IP), a possibly non-isolated charged lepton in the jet, and the presence of a SV. From [66]

In general, this technique of tagging is poor, since any weak decay may lead to soft lepton in the final state; however, used in combination with other taggers, a soft-lepton tagger may significantly increase the performance of the discrimination.

The specific taggers in CMS will be further detailed in Sect. 7.4, while performing the analysis.

5.3.2 Mechanism of Production of b's with QCD Processes

We review the mechanism of production of b quarks.

At TEVATRON, three categories of production of b quarks were identified [67] (illustrated in Fig. 5.4):

Flavour Creation (FCR) where the pair of b quarks is produced in the final state of the hard process (Fig. 5.4a). In the initial state can be pairs of gluons or pairs of light (or charm) quarks; at tree-level, it corresponds to the following diagrams:

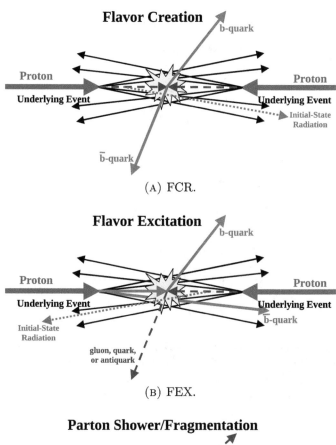

(A) FCR.

(B) FEX.

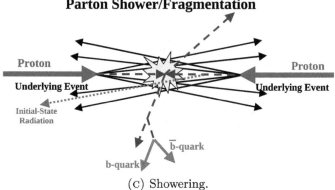

(C) Showering.

Fig. 5.4 Production modes of *b* quarks. The incident protons (or proton-antiproton) are represented horizontally; the hard partons (BBR) are shown with thick (thin) arrows. The dashed (continuous) thick arrows stand for gluons (*b* quarks). Figure modified from [67]

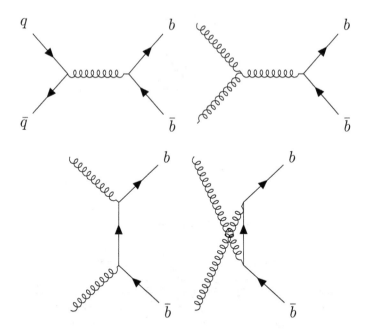

Flavour Excitation (FEX) where a b quark is present in the initial state of the hard process (Fig. 5.4b). A double FEX with two b quarks simultaneously present in the initial state is possible but negligible. The b quark in the initial state is accompanied with a gluon or with a light (or charm) quark; in the 5-flavour scheme the tree-level diagrams are:

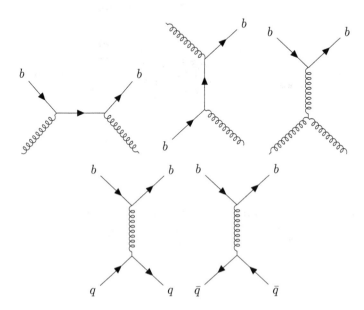

The *b* quark in the initial state must be issued from the evolution of the PDFs. The evolution is usually performed in the massless approximation, as will be the case in the analysis in Part II. For scales $Q^2 \gg m_b^2$, the mass is not expected not play any significant rôle.

Showering a.k.a. Gluon Splitting (GSP), where a pair of *b* quarks is issued from the branching of a gluon in the PS. The most relevant tree-level diagrams are the following:

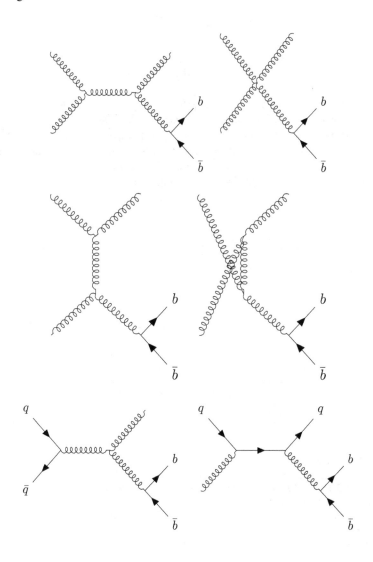

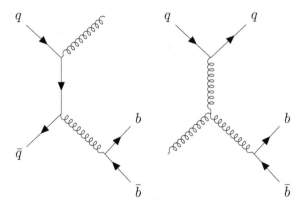

with q for light (or charm) quark. But in principle, GSP can take place on any gluon branch, for instance in combination with FEX. In the PSs, the mass usually only plays a threshold effect: for scales $Q^2 < (2m_b)^2$, no pairs of b quarks can be produced anymore. The contributions from GSP can be understood from the angular separation of $b\bar{b}$ systems; a simulation of the production of B mesons with and without GSP is compared to CMS data at 7 TeV in Fig. 5.5, and illustrates the important contributions from the PS.

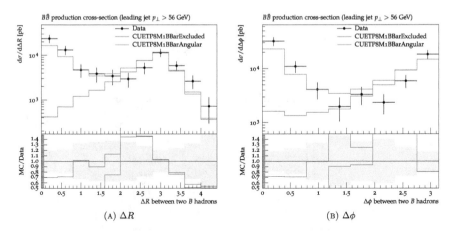

(A) ΔR (B) $\Delta\phi$

Fig. 5.5 The differential $B\bar{B}$ cross section is measured as a function of the opening angle ΔR and $\Delta\phi$ using data collected with the CMS detector during 2010 and corresponding to an integrated luminosity of 3.1 pb^{-1}. The measurement is performed for three different event energy scales, characterized by the transverse momentum of the leading jet in the event (above 56 GeV, [...]). Simulated events are normalised in the region $\Delta R > 2.4$ and $\Delta\phi > \frac{3\pi}{4}$ respectively [68]. The present plots are performed with PYTHIA 8 through the RIVET interface [69]. The blue (red) line corresponds to the situation where gluon splitting in $b\bar{b}$ in the PS is included (excluded)

5.3.3 b *as a Test for QCD*

We briefly review three topics of QCD where *b* quarks could be of interest: the evolution, flavour democracy and tuning.

5.3.3.1 Evolution

As a significant fraction of the *b* jets are issued from the PS and from FEX, various measurements of the *b* jet production can be used to investigate the evolution. For instance, as shown in Fig. 5.6, attempts to use CCFM evolution to describe the inclusive *b* jet production were performed with data from TEVATRON [70, 71]. In addition to inclusive *b* jet measurements, measurements of the angular separation of $b\bar{b}$ pairs can help studying the extra radiations and test CCFM evolution.

5.3.3.2 Flavour Democracy

Since QCD interactions are not sensitive to the flavour, an important verification of QCD consists in comparing the production of jets of different flavours. In regions of the phase space where the masses become negligible, the cross section should be the same for all flavours. This is illustrated in Fig. 5.7, where the partonic cross sections of *b* quark and *t* quark productions are compared in simulation for at $p_T \sim \mathcal{O}(1\,\text{TeV})$; in particular, from 1 TeV, even the mass of the top can be neglected.

Several jet measurements can be repeated for *b* jets. First, the inclusive jet and *b* jet cross section can be compared—this will be presented in Part II of this thesis. In addition, the angular correlations of jets at high multiplicity can also be repeated for *b* jets, as well as the determination of the strong coupling from a ratio of cross section with three and two jets in the final state.

Fig. 5.6 Cross section for $b\bar{b}$ production with $|y^b| < 1$ as a function of $p_T^{\min}$. Shown are the DØ [36] data points, the fixed-order NLO prediction, and the prediction of CASCADE [71]

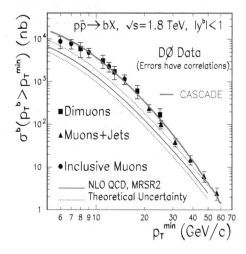

Fig. 5.7 Illustration of flavour democracy with b and t quarks. The cross sections are absolute and have been produced with PYTHIA 8; the plots are obtained with the Rivet interface [69]

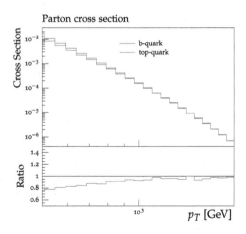

5.3.3.3 Phenomenology

Finally, the phenomenology of the UE, as described in Sect. 2.3, has to be tested, or the parameters of the current tunes need to be verified or refined. For instance, as we saw in Sect. 2.3.3.4, fragmentation requires dedicated parameters for heavy quarks.

In principle, tuning (already introduced in Sect. 4.2.2.2) may be performed with any distributions that is sensitive to the soft activity. Typically, measurements of angular separation of $B\bar{B}$ pairs of hadrons would be crucial to improve the description of hadronisation, which, as we mentioned in Sect. 4.2.2.2, still relies on LEP measurements.

5.3.4 b as a Probe for Other Processes

Although we are not concerned with other SM or BSM processes, the b jets often have a privileged rôle, since they are rather easy to detect. In order to illustrate it, we give here a short review of top physics, Higgs physics and searches.

5.3.4.1 Top Physics

The top quark decays with a lifetime much shorter than the time scale for hadronisation. Thus, the top quark may be studied as a bare quark (i.e. without the complications related to hadronisation in the study of all other flavours). The main channel to measure it is the weak decay:

$$t \to Wb \tag{5.10}$$

It is therefore crucial to know well and detect efficiently b quarks (or B mesons) [72–74]. In addition, the W itself can decay hadronically, potentially giving another b quark.

5.3.4.2 Higgs Physics

The Higgs boson was already mentioned in Chap. 2. Originally, the Higgs boson was discovered using the $H \to \gamma\gamma$ channel [75, 76]:

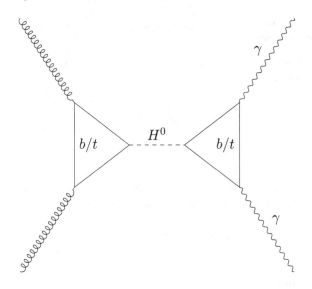

However, from Fig. 5.8, one can see that the $H \to b\bar{b}$ channel has a much larger branching ratio:

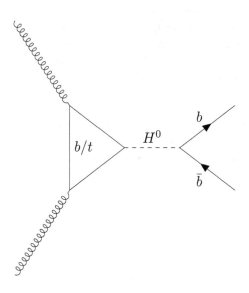

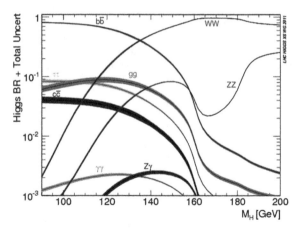

Fig. 5.8 Higgs branching ratios as a function of the mass of the Higgs boson [82]. The Golden channel, involving b's, has an important branching ratio. This plot is historically important since it drove the design of LHC experiments in the quest of the Higgs boson

Despite the much larger branching ratio, there is « only *evidence* » for this decay channel,[14] which is due to overwhelming $b\bar{b}$ background.

Currently, many Higgs analyses rely on the presence of b jets in the final state. Only at CMS, one can cite many analyses [77–81].

5.3.4.3 Searches

b jets are heavily used in searches, either as a background or as part of the signal. In both cases, tagging techniques make b jets privileged objects since they are easy to manipulate, with a rather high efficiency. At CMS, one can mention for instance searches for heavy vector bosons [83], searches for light Higgs bosons [84] or searches for leptoquarks [85]. This thesis does not treat on searches and extensions of the SM; therefore, we shall not detail them.

Conclusion

There can hardly be any doubt about the importance of the beauty quark in physics at LHC: top and Higgs physics, searches, hadronisation, calculations, PDFs. A deep understanding of its properties is therefore essential, and motivates precision measurements. At CMS for LHC Run-II, the b jet production is one of the first observables to investigate.

---

[14]Observation conventionally means at least five sigma of significance.

References

1. Thorndike EH (1999) Bottom quark physics: past, present, future. In: Probing luminous and dark matter. Proceedings, symposium in honor of Adrian Melissinos, Rochester, USA, 24–25 September 1999, pp 127–159. arXiv:hep-ex/0003027 [hep-ex]
2. Appel JA et al (1974) A study of di-lepton production in proton collisions at NAL
3. Fermilab (1977) Discoveries at Fermilab - discovery of the bottom quark. http://lepewwg.web.cern.ch/LEPEWWG
4. Herb SW et al (1977) Observation of a dimuon resonance at 9.5 GeV in 400-GeV proton-nucleus collisions. Phys Rev Lett 39:252–255. https://doi.org/10.1103/PhysRevLett.39.252
5. Fermilab history and archives project. Discovery of the bottom quark, upsilon. https://history.fnal.gov/botqrk.html. Accessed 3 Dec 2017
6. CERN courrier (2017) Revisiting the b revolution. http://cerncourier.com/cws/article/cern/68794
7. Augustin JE et al (1974) Discovery of a narrow resonance in e^+e^- annihilation. Phys Rev Lett 33:1406–1408 [Adv Exp Phys 5:141 (1976)]. https://doi.org/10.1103/PhysRevLett.33.1406
8. Aubert JJ et al (1974) Experimental observation of a heavy particle J. Phys Rev Lett 33:1404–1406. https://doi.org/10.1103/PhysRevLett.33.1404
9. Bjørken BJ, Glashow SL (1964) Elementary particles and SU(4). Phys Lett 11(3):255–257. https://doi.org/10.1016/0031-9163(64)90433-0. ISSN:0031-9163
10. Glashow SL, Iliopoulos J, Maiani L (1970) Weak interactions with lepton-hadron symmetry. Phys Rev D 2:1285–1292. https://doi.org/10.1103/PhysRevD.2.1285
11. Appelquist T, Politzer HD (1975) Heavy quarks and e^+e^- annihilation. Phys Rev Lett 34:43–45. https://doi.org/10.1103/PhysRevLett.34.43
12. Kobayashi M, Maskawa T (1973) CP-violation in the renormalizable theory of weak interaction. Prog Theor Phys 49(2):652–657. https://doi.org/10.1143/PTP.49.652
13. Fabjan CW, McCubbin N (2004) Physics at the CERN intersecting storage rings (ISR) 1978–1983. Phys Rep 403–404, 165–175. https://doi.org/10.1016/j.physrep.2004.08.018, http://www.sciencedirect.com/science/article/pii/S0370157304003308. ISSN:0370-1573
14. Cobb JH et al (1977) The cross section for the production of massive electron pairs and the Υ (9.5 GeV) in proton-proton collisions at the CERN ISR. Phys Lett B 72(2):273–277. https://doi.org/10.1016/0370-2693(77)90720-1. ISSN:0370-2693
15. Berger Ch et al (1978) Observation of a narrow resonance formed in e+e annihilation at 9.46 GeV. Phys Lett B 76(2):243–245. https://doi.org/10.1016/0370-2693(78)90287-3. ISSN:0370-2693
16. Darden CW et al (1978) Observation of a narrow resonance at 9.46 GeV in electron - positron annihilations. Phys Lett B 76(2):246–248. https://doi.org/10.1016/0370-2693(78)90288-5. ISSN:0370-2693
17. Darden CW et al (1979) Study of the (9.46) meson in electron-positron annihilations. Phys Lett B 80(4):419–422. https://doi.org/10.1016/0370-2693(79)91204-8. ISSN:0370-2693
18. Bienlein JK et al (1978) Observation of a narrow resonance at 10.02 GeV in e+e annihilations. Phys Lett B 78(2):360–363. https://doi.org/10.1016/0370-2693(78)90040-0. ISSN:0370-2693
19. Eichten E et al (1980) Charmonium: comparison with experiment. Phys Rev D 21:203. https://doi.org/10.1103/PhysRevD.21.203
20. Franzini P, Lee-Franzini J (1982) Upsilon physics at CESR. Phys Rep 81(3):239–291. https://doi.org/10.1016/0370-1573(82)90027-8. ISSN:0370-1573
21. Andrews D et al (1980) Observation of three upsilon states. Phys Rev Lett 44:1108–1111. https://doi.org/10.1103/PhysRevLett.44.1108
22. Andrews D et al (1980) Observation of a fourth upsilon state in e^+e^- annihilations. Phys Rev Lett 45:219–221. https://doi.org/10.1103/PhysRevLett.45.219
23. Finocchiaro G et al (1980) Observation of the upsilon at CESR. Phys Rev Lett 45:222. https://doi.org/10.1103/PhysRevLett.45.222
24. Tanabashi M et al (2018) Review of particle physics. Phys Rev D 98(3):030001

25. Besson D, Skwarnicki T (1993) v spectroscopy. Ann Rev Nucl Part Sci 43:333–378. https://doi.org/10.1146/annurev.ns.43.120193.002001

26. Abe T et al (2010) Belle II technical design report. arXiv:1011.0352 [physics.ins-det]

27. Cassel DG (2007) CLEO B physics. In: ARGUS Fest, 20 years of B meson mixing 1987–2007. Proceedings, ARGUS-symposium, DESY, Hamburg, Germany, 9 November 2007. https://doi.org/10.3204/DESY-PROC-2008-01/e307

28. Lee-Franzini J (1998) Hidden and open beauty in CUSB. AIP Conf Proc 424(1):85–96. https://doi.org/10.1063/1.55115, arXiv:hep-ex/9709025 [hep-ex]

29. Albrecht H et al (1989) Argus: a universal detector at DORIS II. Nucl Instrum Methods Phys Res Sect A: Accel Spectrom Detect Assoc Equip 275(1):1–48

30. Albajar C et al (1987) Beauty production at the CERN proton-antiproton collider. In: 186, pp 237–246

31. Albajar C et al (1987) Beauty production at the CERN proton - anti-proton collider. 1. Phys Lett B 186:237–246. https://doi.org/10.1016/0370-2693(87)90287-5

32. Albajar C et al (1988) Measurement of the bottom quark production cross-section in proton - anti-proton collisions at s**(1/2) = 0.63−TeV. Phys Lett B 213:405 (473 (1988)). https://doi.org/10.1016/0370-2693(88)91785-6

33. Abazov VM et al (2010) b-jet identification in the D0 experiment. Nucl Instrum Methods A 620:490–517. https://doi.org/10.1016/j.nima.2010.03.118, arXiv:1002.4224 [hep-ex]

34. Abe F et al (1996) Inclusive jet cross section in $\bar{p}p$ collisions at $\sqrt{s} = 1.8$ TeV. Phys Rev Lett 77:438–443. https://doi.org/10.1103/PhysRevLett.77.438, arXiv:hep-ex/9601008 [hep-ex]

35. Abbott B et al (2000) Cross section for b jet production in $\bar{p}p$ collisions at $\sqrt{s} = 1.8$ TeV. Phys Rev Lett 85:5068–5073. https://doi.org/10.1103/PhysRevLett.85.5068, arXiv:hep-ex/0008021 [hep-ex]

36. Abbott B et al (2000) Cross section for b-jet production in pp collisions at $\sqrt{s} = 1.8$ TeV. Phys Rev Lett 85(24):5068

37. Abe F et al (1993) Measurement of bottom quark production in 1.8 TeV $p\bar{p}$ collisions using semileptonic decay muons. Phys Rev Lett 71:2396–2400. https://doi.org/10.1103/PhysRevLett.71.2396

38. Abe F et al (1993) Measurement of the bottom quark production cross-section using semileptonic decay electrons in $p\bar{p}$ collisions at $\sqrt{s} = 1.8$ TeV. Phys Rev Lett 71:500–504. https://doi.org/10.1103/PhysRevLett.71.500

39. Abe F et al (1995) Measurement of the B meson differential cross-section, $d\sigma/dp_T$, in $p\bar{p}$ collisions at $\sqrt{s} = 1.8$ TeV. Phys Rev Lett 75:1451–1455. https://doi.org/10.1103/PhysRevLett.75.1451, arXiv:hep-ex/9503013 [hep-ex]

40. Behnke O, Geiser A, Lisovyi M (2015) Charm, beauty and top at HERA. Prog Part Nucl Phys 84:1–72. https://doi.org/10.1016/j.ppnp.2015.06.002, arXiv:1506.07519 [hep-ex]

41. Chekanov S et al (2009) Measurement of beauty production from dimuon events at HERA. JHEP 02:032. https://doi.org/10.1088/1126-6708/2009/02/032, arXiv:0811.0894 [hep-ex]

42. Rozen Y, Pierre F, Calderini G (2000) B physics at LEP. Nucl Instrum Methods Phys Res Sect A: Accel Spectrom Detect Assoc Equip 446(1):37–52. https://doi.org/10.1016/S0168-9002(00)00016-4. ISSN:0168-9002

43. Rozen Y (2003) The legacy of LEP B program. In: Nuclear physics B - proceedings supplements 120. Supplement C. Proceedings of the 8th international conference on B-physics at hadron machines, pp 83–90. https://doi.org/10.1016/S0920-5632(03)01885-1. ISSN:0920-5632

44. Abdallah J et al (2011) A study of the b-quark fragmentation function with the DELPHI detector at LEP I and an averaged distribution obtained at the Z Pole. Eur Phys J C 71:1557. https://doi.org/10.1140/epjc/s10052-011-1557-x, arXiv:1102.4748 [hep-ex]

45. Abe K et al (2004) Observation of large CP violation and evidence for direct CP violation in $B^0 \to \pi^+\pi^-$ decays. Phys Rev Lett 93:021601. https://doi.org/10.1103/PhysRevLett.93.021601

46. Chao Y et al (2004) Evidence for direct CP violation in $B^0 \to K^+\pi^-$ decays. Phys Rev Lett 93:191802. https://doi.org/10.1103/PhysRevLett.93.191802

47. Gershon T, Needham M (2015) Heavy flavour physics at the LHC. Comptes Rendus Phys 16:435–447. https://doi.org/10.1016/j.crhy.2015.04.001, arXiv:1408.0403 [hep-ex]
48. Chatrchyan S et al (2012) Inclusive b-jet production in pp collisions at $\sqrt{s} = 7$ TeV. JHEP 04:084. https://doi.org/10.1007/JHEP04(2012)084, arXiv:1202.4617 [hep-ex]
49. Khachatryan V et al (2011) Measurement of the B^+ production cross section in pp collisions at $\sqrt{s} = 7$ TeV. Phys Rev Lett 106:112001. https://doi.org/10.1103/PhysRevLett.106.112001, arXiv:1101.0131 [hep-ex]
50. Khachatryan V et al (2011) Inclusive b-hadron production cross section with muons in pp collisions at $\sqrt{s} = 7$ TeV. JHEP 03:090. https://doi.org/10.1007/JHEP03(2011)090, arXiv:1101.3512 [hep-ex]
51. Chatrchyan S et al (2011) Measurement of the B^0 production cross section in pp collisions at $\sqrt{s} = 7$ TeV. Phys Rev Lett 106:252001. https://doi.org/10.1103/PhysRevLett.106.252001, arXiv:1104.2892 [hep-ex]
52. Chatrchyan S et al (2011) Measurement of the strange B meson production cross section with J/Psi ϕ decays in pp collisions at $\sqrt{s} = 7$ TeV. Phys Rev D 84:052008. https://doi.org/10.1103/PhysRevD.84.052008, arXiv:1106.4048 [hep-ex]
53. Aad G et al (2011) Measurement of the inclusive and dijet cross-sections of b^- jets in pp collisions at $\sqrt{s} = 7$ TeV with the ATLAS detector. Eur Phys J C 71:1846. https://doi.org/10.1140/epjc/s10052-011-1846-4, arXiv:1109.6833 [hep-ex]
54. Aad G et al (2012) Measurement of the b-hadron production cross section using decays to $D^*\mu^- X$ final states in pp collisions at $\sqrt{s} = 7$ TeV with the ATLAS detector. Nucl Phys B 864:341–381. https://doi.org/10.1016/j.nuclphysb.2012.07.009, arXiv:1206.3122 [hep-ex]
55. Aad G et al (2013) Measurement of the differential cross-section of B^+ meson production in pp collisions at $\sqrt{s} = 7$ TeV at ATLAS. JHEP 10:042. https://doi.org/10.1007/JHEP10(2013)042, arXiv:1307.0126 [hep-ex]
56. Aaij R et al (2010) Measurement of $\sigma(pp \rightarrow b\bar{b}X)$ at $\sqrt{s} = 7$ TeV in the forward region. Phys Lett B 694:209–216. https://doi.org/10.1016/j.physletb.2010.10.010, arXiv:1009.2731 [hep-ex]
57. Aaij R et al (2012) Measurement of the $B^\pm$ production cross-section in pp collisions at $\sqrt{s} = 7$ TeV. JHEP 04:093. https://doi.org/10.1007/JHEP04(2012)093, arXiv:1202.4812 [hep-ex]
58. Aaij R et al (2013) Measurement of B meson production cross-sections in proton-proton collisions at $\sqrt{s} = 7$ TeV. JHEP 08:117. https://doi.org/10.1007/JHEP08(2013)117, arXiv:1306.3663 [hep-ex]
59. Aaij R et al (2015) Measurement of B_c^+ production in proton-proton collisions at $\sqrt{s} = 8$ TeV. Phys Rev Lett 114:132001. https://doi.org/10.1103/PhysRevLett.114.132001, arXiv:1411.2943 [hep-ex]
60. Alessandro Grelli and the ALICE Collaboration (2011) Heavy flavour physics with the ALICE detector at the CERN-LHC. J Phys: Conf Ser 316(1):012025. http://stacks.iop.org/1742-6596/316/i=1/a=012025
61. Geiser A, Definition of stable hadron in ZEUS. Private communication
62. Jung H, Definition of stable hadron in H1. Private communication
63. Scodellaro L (2017) b tagging in ATLAS and CMS. In: 5th large hadron collider physics conference (LHCP 2017) Shanghai, China, 15–20 May 2017. https://inspirehep.net/record/1621595/files/arXiv:1709.01290.pdf, arXiv:1709.01290 [hep-ex]
64. Aad G et al (2016) Performance of b-jet identification in the ATLAS experiment. JINST 11(04):P04008. https://doi.org/10.1088/1748-0221/11/04/P04008, arXiv:1512.01094 [hep-ex]
65. Chatrchyan S et al (2013) Identification of b-quark jets with the CMS experiment. JINST 8:P04013. https://doi.org/10.1088/1748-0221/8/04/P04013, arXiv:1211.4462 [hep-ex]
66. CMS Collaboration (2016) Identification of b quark jets at the CMS experiment in the LHC Run 2
67. Field RD (2002) The sources of b quarks at the Tevatron and their correlations. Phys Rev D 65:094006. https://doi.org/10.1103/PhysRevD.65.094006, arXiv:hep-ph/0201112 [hep-ex]

68. Khachatryan V et al (2011) Measurement of dijet angular distributions and search for quark compositeness in pp collisions at $sqrts$ = 7 TeV. Phys Rev Lett 106:201804. https://doi.org/10.1103/PhysRevLett.106.201804, arXiv:1102.2020 [hep-ex]
69. Buckley A et al (2013) Rivet user manual. Comput Phys Commun 184:2803–2819. https://doi.org/10.1016/j.cpc.2013.05.021, arXiv:1003.0694 [hep-ex]
70. Jung H (2002) Heavy quark production at the TEVATRON and HERA using k_t factorization with CCFM evolution. Phys Rev D 65:034015. https://doi.org/10.1103/PhysRevD.65.034015, arXiv:hep-ph/0110034 [hep-ex]
71. Jung H (2002) Unintegrated parton densities applied to heavy quark production in the CCFM approach. J Phys G 28:971–982. https://doi.org/10.1088/0954-3899/28/5/320, arXiv:hep-ph/0109146 [hep-ex]
72. Sirunyan AM et al (2017) Cross section measurement of t-channel single top quark production in pp collisions at $\sqrt{s}$ = 13 TeV. Phys Lett B 772:752–776. https://doi.org/10.1016/j.physletb.2017.07.047, arXiv:1610.00678 [hep-ex]
73. Sirunyan AM et al (2017) Measurement of the $t\bar{t}$ production cross section using events with one lepton and at least one jet in pp collisions at $\sqrt{s}$ = 13 TeV. JHEP 09:051. https://doi.org/10.1007/JHEP09(2017)051, arXiv:1701.06228 [hep-ex]
74. Khachatryan V et al (2017) Measurement of differential cross sections for top quark pair production using the lepton+jets final state in proton-proton collisions at 13 TeV. Phys Rev D 95(9):092001. https://doi.org/10.1103/PhysRevD.95.092001, arXiv:1610.04191 [hep-ex]
75. Chatrchyan S et al (2012) Search for the standard model Higgs boson decaying to bottom quarks in pp collisions at $\sqrt{s}$ = 7 TeV. Phys Lett B 710:284–306. https://doi.org/10.1016/j.physletb.2012.02.085, arXiv:1202.4195 [hep-ex]
76. Aaboud M et al (2017) Evidence for the $H \rightarrow b\bar{b}$ decay with the ATLAS detector. JHEP 12:024. https://doi.org/10.1007/JHEP12(2017)024, arXiv:1708.03299 [hep-ex]
77. CMS Collaboration (2016) Search for resonant pair production of Higgs bosons decaying to two bottom quark-antiquark pairs in proton-proton collisions at 13 TeV
78. Search for resonant pair production of Higgs bosons decaying to bottom quark-antiquark pairs in proton-proton collisions at 13 TeV. Technical report CMS-PAS-HIG-17-009. Geneva: CERN (2017). http://cds.cern.ch/record/2292044
79. CMS Collaboration (2017) Evidence for the decay of the Higgs boson to bottom quarks
80. Search for Higgs boson pair production in the final state containing two photons and two bottom quarks in proton-proton collisions at $\sqrt{s}$ = 13 TeV. Technical report CMS-PAS-HIG-17-008. Geneva: CERN (2017). http://cds.cern.ch/record/2273383
81. Inclusive search for the standard model Higgs boson produced in pp collisions at $\sqrt{s}$ = 13 TeV using $H \rightarrow$ bb decays. Technical report CMS-PAS-HIG-17-010. Geneva: CERN (2017). http://cds.cern.ch/record/2266164
82. Denner A et al (2011) Standard model Higgs-boson branching ratios with uncertainties. Eur Phys J C 71:1753. https://doi.org/10.1140/epjc/s10052-011-1753-8, arXiv:1107.5909 [hep-ex]
83. Chatrchyan S et al (2013) Search for a W' boson decaying to a bottom quark and a top quark in pp collisions at $\sqrt{s}$ = 7 TeV. Phys Lett B 718:1229–1251. https://doi.org/10.1016/j.physletb.2012.12.008, arXiv:1208.0956 [hep-ex]
84. Khachatryan V et al (2015) Search for neutral MSSM Higgs bosons decaying into a pair of bottom quarks. JHEP 11:071. https://doi.org/10.1007/JHEP11(2015)071, arXiv:1506.08329 [hep-ex]
85. Chatrchyan S et al (2013) Search for pair production of third-generation leptoquarks and top squarks in pp collisions at $\sqrt{s}$ = 7 TeV. Phys Rev Lett 110(8):081801. https://doi.org/10.1103/PhysRevLett.110.081801, arXiv:1210.5629 [hep-ex]

Part II
Physics Analysis

Chapter 6
General Strategy and Outline of the Analysis

THE GENERAL STRATEGY FOR THE MEASUREMENT of the double differential cross section of inclusive b jet production using pp collisions with $\sqrt{s} = 13\,\text{TeV}$ recorded with the CMS detector is presented here.

First, the signal of the inclusive b jet production is studied in simulation. Second, previous inclusive b jet analyses at LHC are reviewed. Lastly, following on the conclusions from studies in simulation and previous analyses, the strategy of the measurement is outlined.

6.1 Monte Carlo Studies

In this section, the contributions from QCD and from other processes is described from the simulation of the ME and of the PS with PYTHIA 8 in LHC conditions with NNPDF 2.3 set [1] (shown and compared to other sets in Fig. 6.3).

6.1.1 Contributions from QCD Processes

The QCD diagrams of the production of b quarks were discussed in Sect. 5.3.2.

First, the contributions from b jets to inclusive jet cross section is shown in Fig. 6.1. In the top row of this figure, the inclusive jet production is described in terms of partons in the initial state of the hard process: two quarks (blue), two gluons (green) or a quark-gluon pair (red). The contribution from b jets is shown in darker shades in the top row and alone in the bottom row; the contribution from gg in the initial state is significantly reduced in comparison to the two other initial states.

In Fig. 6.2, the contributions to inclusive b jets cross section are shown in terms of FCR, FEX, GSP. The FCR contributions are the less important one with around

© Springer Nature Switzerland AG 2019
P. Connor, *Inclusive b Jet Production in Proton-Proton Collisions*, Springer Theses,
https://doi.org/10.1007/978-3-030-34383-5_6

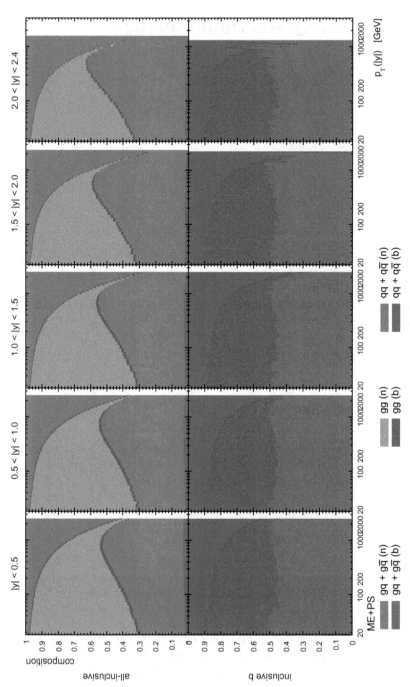

Fig. 6.1 Contributions to inclusive jet (top) and inclusive b jet (bottom) cross section at parton level as a function of the transverse momentum of the jets, as described by PYTHIA 8 (ME+PS) with NNPDF 2.3 set [1]. Blue (green; red) corresponds to processes with two quarks (two gluons; one gluon and one quark) in the initial state; the dark (light) shade corresponds to b jets (non-b jets, or n jets). The five columns correspond to rapidity bins

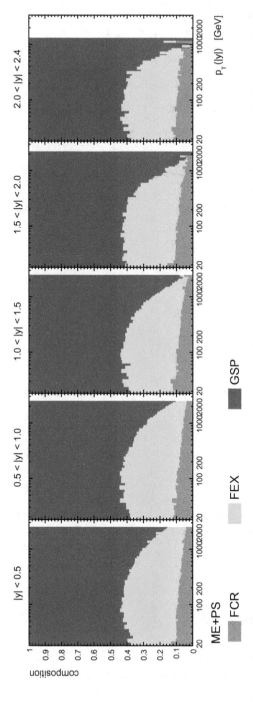

Fig. 6.2 Contributions to inclusive *b* jet cross section at parton level as a function of the transverse momentum of the jets, as described by PYTHIA 8 (ME+PS) with NNPDF 2.3 set [1]. Violet (cyan; yellow) corresponds to FCR (FEX; GSP). The five columns correspond to rapidity bins

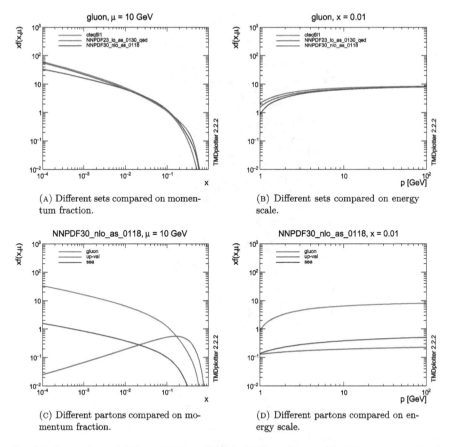

Fig. 6.3 Comparison of different PDF sets [1–3] in LHC conditions with different sets (top) and partons (bottom). The subfigures on the RHS (LHS) show the dependence on the scale μ (on the momentum fraction x). The curves come in pairs and show the uncertainties on the PDF sets. Performed with TMDplotter 2.2.0 [4, 5]

10% of the signal, while the GSP contributions are getting more important at higher transverse momentum of the b jets and represent 50–80%.

From this studies in simulation, one concludes that the contribution from PS is important to describe the inclusive b jet production.

6.1.2   Other Processes

According to the SM, quarks have both strong and, in a lesser extent, electroweak interactions. In this subsection, we investigate the different channels contributing to the inclusive b jet production.

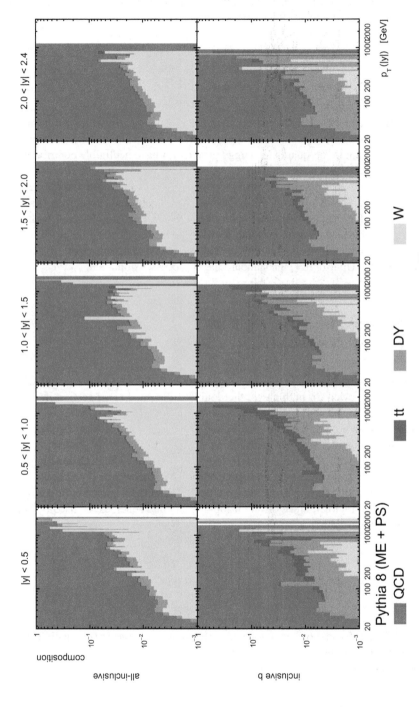

Fig. 6.4 Contributions from the most important modes of production of b jets in pp collisions at $\sqrt{s} = 13$ TeV, obtained from PYTHIA 8 (with CUETP8M1 and NNPDF 2 . 3 [1])

In addition to the QCD processes, the processes listed below can make significant contributions to the inclusive b jet cross section.

$$t\bar{t} \rightarrow b\bar{b}X \hspace{4cm} (t\bar{t}) \hspace{3cm} (6.1)$$

$$W \rightarrow bX \hspace{4cm} (W) \hspace{3cm} (6.2)$$

$$Z \rightarrow b\bar{b}X \hspace{4cm} (DY) \hspace{3cm} (6.3)$$

In principle, $t\bar{t}$ should be considered as part of QCD: however, given the specific properties of the top quark, it is considered separately here. Other signals like electroweak production of single top quarks, or Higgs production have negligible contributions and are not considered.

The contributions from the different processes are shown in Fig. 6.4 as a function of the transverse momentum of the b jets and of the (absolute) rapidity at parton level. The predictions are calculated with PYTHIA 8 and jets are clustered with the anti-k_T algorithm with cone size radius $R = 0.4$.

Here, the predictions do not include any simulation of the detector (like resolution and efficiency). Up to 1 TeV, the standard QCD processes dominate.

6.2 Previous Measurements of the Inclusive b Jet Production at the LHC

The latest measurements of inclusive b jet production in ATLAS [6] and CMS [7] are now reviewed. In both cases, the double differential cross section in transverse momentum and absolute rapidity was measured.

Both collaborations performed the analysis once with soft-lepton tagging and once with SV tagging; the four measurements are performed with anti-k_T jet clustering algorithm with $R = 0.5$ and corrected to parton level. The measurements are represented and compared with one another and with theoretical predictions from MC@NLO in Fig. 6.5; however, the measurements can only be compared in a single, inclusive bin of rapidity, since different rapidity binning schemes are used by each collaboration.

6.2.1 ATLAS at 7 TeV

The comparison of the ATLAS [6] measurements to NLO parton-level predictions is shown in Figs. 6.5 and 6.6. The muon and vertex based analyses are compatible with one another within the systematic uncertainties. POWHEG+PYTHIA (green) shows a better agreement than MC@NLO+HERWIG (red).

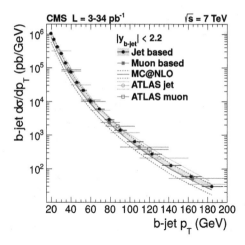

Fig. 6.5 Absolute differential cross section in transverse momentum at $\sqrt{s} = 7$ TeV of inclusive the *b*-quark jet production, measured in ATLAS (green and purple) and CMS (yellow and blue) and predicted with MC@NLO [7]. Four series of data points are shown, since both collaboration performed the analysis once with soft-lepton tagging (*muon*) and once with secondary-vertex tagging (*jet*). The CMS luminosity is $3\,\text{pb}^{-1}$ ($34\,\text{pb}^{-1}$) for the muon (jet) analysis, while ATLAS luminosity is $34\,\text{pb}^{-1}$ in both cases

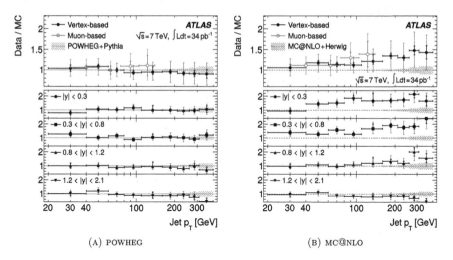

(A) POWHEG   (B) MC@NLO

Fig. 6.6 Ratio of the measured cross-sections to the theory predictions of POWHEG and MC@NLO. In the region where the lifetime-based measurement overlaps with the muon p_T^{rel} measurement both results are shown. The top plot shows the full rapidity acceptance, while the four smaller plots show the comparison for each of the rapidity ranges separately. The data points show both the statistical uncertainty (dark colour) and the combination of the statistical and systematic uncertainty (light colour). The shaded regions around the theoretical predictions reflect the statistical uncertainty only. [...] [6]

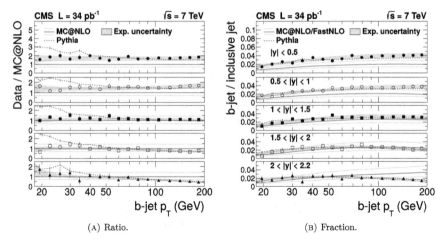

(A) Ratio.   (B) Fraction.

Fig. 6.7 Measured *b* jet cross section from the jet analysis (yellow band and error bars for respectively systematics and statistical uncertainties), compared to the MC@NLO calculation (red lines, nominal and uncertainties are respectively plain and dotted) and to PYTHIA prediction (no uncertainty shown) [7]

6.2.2   CMS at 7 TeV

The measurements at CMS [7] are shown in Figs. 6.5, 6.6 and 6.7. In addition to the ratio of the cross sections of the inclusive *b* jet production in data and simulation, the fraction of *b* jets in the inclusive jet production is also given.

Both measurements were published for $18\,\text{GeV} < p_T < 200\,\text{GeV}$. The conditions of low pile-up allowed to reach low transverse momentum. However, it is interesting to note that the upper boundary is not due to limited statistics but due to the reliability in the *b* tagging [8].

PYTHIA 6 (blue) exhibits a better description of the ratio with data (Fig. 6.6a), while MC@NLO (red) describes better the fraction of *b* jets in data (Fig. 6.7b).

6.3   Strategy

The aim of the present analysis is to perform a new *precision measurement* of the inclusive *b* jet cross section and fraction to the inclusive jet cross section, similarly to the measurements at $\sqrt{s} = 7\,\text{TeV}$ presented in Sect. 6.2. With the improvement in *b* tagging techniques from Run-I to Run-II and thanks to samples of much larger statistics, the measurable phase space is extended to the TeV scale.

The analysis will be conducted in two steps: first at detector level (Chap. 7), then at particle level (Chap. 8):

- At detector level, the effect of the detector is investigated in the simulation. Thanks to comparisons with data, corrections are applied to simulation to compensate for imperfect modelling. Moreover, the quality of the data is ensured and checked to be constant over time.
- Data distribution are then corrected to particle level (more precisely, to hadron level). The procedure of correction from detector level to particle level is called *unfolding*.

The procedure of *unfolding* is crucial and underlies the whole analysis: the analysis at detector level ensures that the simulation describes optimally the data; then all detector and reconstruction effects are corrected together through the unfolding procedure.

Only after unfolding, conclusions from comparisons to theoretical predictions are drawn (Chap. 9). While comparisons to theoretical predictions (folded with the simulation of the detector) could be performed at detector level, unfolding is essential to allow comparisons with other measurements. It is also crucial for the extraction of parameters (e.g. α_S or tuning parameters) or of PDF.

In this section, we first describe the phase space and the selection (Sect. 6.3.1), as well as the data and simulation samples that will be used to conduct the analysis (Sect. 6.3.2). Then we outline the different steps of the analysis in more detail (Sects. 6.3.3–6.3.4). Finally, we discuss the theoretical predictions to which the measurement will be compared.

6.3.1 Phase Space Definition and Selection

The phase space definition and selection is summarised in Table 6.1.

- The phase space will be essentially limited by the pile-up at low transverse momentum and by the tracker acceptance for the rapidity coverage (including muon chambers).
- The selection of the events at detector level is motivated by the vertexing performance. Fake vertices due to a bad resolution and high contamination from pile-up are discarded. The standard definition of *good* event is performed centrally in the collaboration and corresponds to an event containing at least one *good* Primary Vertex (PV), defined as follow:

  - at least four tracks in the vertex fit;
  - the vertex position must satisfy $|z_{PV}| < 24$ cm around the IP, along the beam pipe and $|r_{PV}| < 2$ cm in the transverse plane to the beam pipe.

The jet reconstruction at detector level aims at eliminating jets contaminated by pile-up; the *jet ID* is based on the stable particles entering the composition of the jet. In order to apply the calibration on jets, it is important to ensure that the jets are correctly described by the simulation.

Table 6.1 Phase space and selection of the analysis. The reconstruction criteria are centrally defined at CMS; the kinematics are restricted by the pile-up and by the tracker coverage; the b tagging is also define centrally

	Selection
	Selection
Reconstruction	Good PV
	Jet tight ID
Kinematics	$p_T^{\text{jet}} > 74\,\text{GeV}$
	$|y^{\text{jet}}| < 2.4$
b tagging	Tight selection

– The b-tagging variable that shall be used by default in this analysis is primarily based on the detection of a Secondary Vertex (SV), namely the CSVv2 tagger [9– 11]. At the time when the analysis was started, the choice for the CSVv2 algorithm was motivated by the fact that it was the best compromise between the size of the uncertainties and coverage of the calibration (i.e. up to $p_T = 1\,\text{TeV}$).

All the elements of the selection will be described in further details in Chap. 7. In general, since the analysis consists in a precision measurement, the tightest possible selection is considered.

The discriminant is only applied on all jets to separate into non-b-tagged and b-tagged jets; in the end, the b jet cross section is extracted together with a non-b-jet cross section from the b-tagged and non-b-tagged cross sections.

Convention

In order to ease the discussions throughout the analysis, we use b and n ($\hat{b}$ and $\hat{n}$) for b true and non-b-true (b tagged and non-b-tagged) jets; even if the hadron definition is used, we shall use the convention at CMS and write b jet.

6.3.2   Samples

In this section, we give a description of the data and simulation samples.

6.3.2.1   Data

The data files are obtained after a long chain of certification, calibration and reconstruction that is not described here. The data taking is divided in different eras of various luminosities; the eras and respective luminosities are shown in Table 6.2. The different eras (or runs) correspond to different phases of data taking, and their exact definition is usually not relevant at the level of physics analyses. However, the calibration may be non-constant over time; in particular, the last era (RunH) was

Table 6.2 The luminosities are shown per era, as well as their respective contribution to the total sample in terms of fraction. The number of events corresponds to the number of triggered events

Era	$\mathcal{L}/fb^{-1}$	Fraction
Run2016BCD	12.498	0.355
Run2016EFearly	6.589	0.187
Run2016FlateG	7.884	0.224
Run2016H	8.208	0.233
TOTAL	35.179	1.000

Table 6.3 Simulation samples used to perform the analysis

Generator	PDF	ME	UE
PYTHIA 8	NNPDF 2.3 [1]	LO hard-coded $2 \rightarrow 2$	CUETP8M1
MADGRAPH	NNPDF 2.3 [1]	LO automated $2 \rightarrow 2, 3, 4$	CUETP8M1
HERWIG++	CTEQ6L1 [2]	LO hard-coded $2 \rightarrow 2$	CUETHppS1

reconstructed with a rougher calibration.[1] In the analysis, we shall explicitly check that there is no dependence over time anymore after calibration.

6.3.2.2 Simulation

Physics generators were already introduced in Sect. 4.2.2.1; here, we only describe details specific to samples used in the present analysis. Essentially, three generators will be considered: PYTHIA 8, MADGRAPH (interfaced with PYTHIA 8 for the PS, MPI and hadronisation), and HERWIG++. The simulations are compared in Table 6.3, and the PDF sets are shown in Fig. 6.3. Each sample contains a large number of event records at detector and particle levels; the description of the detector will be constructed by performing *matchings* between the two levels. In addition, all the samples include a simulation of the pile-up in data.

As a conclusion from Sect. 6.1 according to which QCD processes clearly dominate, it seems sufficient to consider only QCD processes in the simulation throughout the analysis. It is important here to stress that these samples will primarily be used to investigate the reconstruction, but not to draw any conclusions on physics. Whether other processes should be added while performing physics comparison depends on the resolution at high transverse momentum, but for the analysis itself, especially the unfolding, only QCD processes are considered.

---

[1]The reason for this is purely technical. At CMS, the size of the samples is such that the reconstruction with finer calibration can only be run once every few months. In particular, at the time of writing of the present thesis, one year after the end of the data taking, no fine calibration of Run2016H was yet available.

Given the steeply falling character of the p_T spectrum, different strategies have been followed to provide sufficient statistics over the full phase space:

PYTHIA 8   The sample is generated in slices of $\hat{p}_T$ (already mentioned in Appendix 4.A). The slices may be seen in Fig. 6.8a and the number of events per slice is summarised in Table 6.4a.

MADGRAPH   Similarly, the sample is generated in slices of the scalar sum of the transverse momenta of the outgoing partons in the ME:

$$H_T = \sum_{\text{jets}} p_T^{\text{jet}} \tag{6.4}$$

The slices may be seen in Fig. 6.8b and the number of events per slice is summarised in Table 6.4b.

HERWIG++   Finally, the HERWIG++ sample is a smallest sample; the events are generated with a $\hat{p}_T$ uniformly distributed from 15 to 7000 GeV; the sample contains 9573,938 events; the cross section is $1667 \cdot 10^6$ pb.

The HERWIG++ sample is a small sample with a slightly older calibration; its particular interest comes from the fact that it is generated completely independently, whereas the MADGRAPH sample uses the PYTHIA interface. Moreover, in addition to being a small sample, since the b calibration and Jet Energy Corrections (JECs) are derived for PYTHIA 8, we only use HERWIG++ as a cross-check in at the detector level.

6.3.3   Analysis at Detector Level

The analysis at detector level consists in checking in detail the data and simulation samples, and to correct the latter in order to improve the description of the former. The inclusive b-tagged jet and inclusive jet productions are analysed simultaneously, in order to understand possible biases due to b tagging.

Formally, the double differential cross section in transverse momentum and rapidity can be written as follows:

$$\frac{\mathrm{d}^2 \sigma_{(b)}^{\text{rec}}}{\mathrm{d}p_T \mathrm{d}|y|} = \frac{N_{(b)}}{\epsilon \, \mathcal{L} \, \Delta p_T \, \Delta |y|} \tag{6.5}$$

where

- p_T (y) stands for the transverse momentum (rapidity),
- N ($N_{\hat{b}}$) stands for the count of the jets (b-tagged jets),
- $\mathcal{L}$ stands for the luminosity,
- $\epsilon(p_T, |y|)$ stands for the efficiencies (trigger, tracking, etc.),
- and Δp_T and Δy stand for the bin widths.

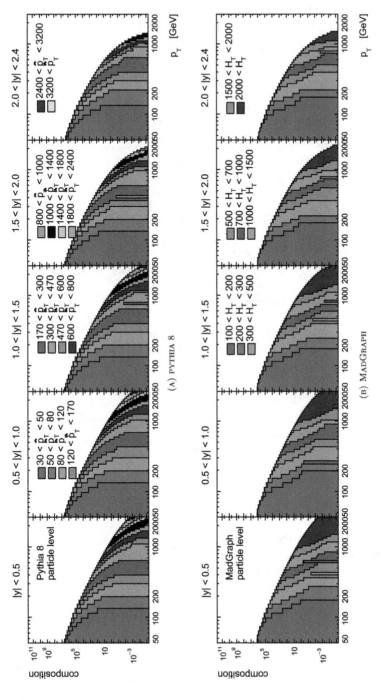

Fig. 6.8 Contributions to the p_T spectrum of the different slices at generator level

Table 6.4 Description of the sub-samples in PYTHIA 8 and MADGRAPH. The number of events and the cross section is given for each slice of $\hat{p}_T$ or H_T

$\hat{p}_T$ slice		#events	σ/pb
30	50	9699558	140932000
50	80	9948791	19204300
80	120	7742665	2762530
120	170	5748730	471100
170	300	7838066	117276
300	470	11701816	7823
470	600	3959986	648.2
600	800	9628335	186.9
800	1000	11915305	32.293
1000	1400	6992746	9.4183
1400	1800	2477018	0.84265
1800	2400	1584378	0.114943
2400	3200	596904	0.00682981
3200	∞	391735	0.000165445

(a) PYTHIA

H_T slice		#events	σ/pb
100	200	69031923	27540000
200	300	18847246	1717000
300	500	17035890	351300
500	700	37645465	31630
700	1000	15534009	6802
1000	1500	4850746	1206
1500	2000	3970819	120.4
2000	∞	1967899	25.25

(b) MADGRAPH

First, the data acquisition is detailed. Given the steeply falling character of the spectrum, the sample is split into different intervals of transverse momentum of the leading jet, corresponding to different acquisition rates (or *trigger rates*); the normalisation and combination of these different regions of the phase space is presented.

Then, various procedures are applied in order to improve the description of the data, and hence their agreement. Important aspects to treat are the pile-up simulation, the jet energy scale and resolution, and the calibration of b tagging. An estimation of each of these effects is given, together with their systematic uncertainties, without necessarily attempting to correct for them: this belongs to the unfolding procedure. Especially, at high transverse momentum, we shall see that large differences related to the b tagging in simulation and in data will be observed, together with a low tagging efficiency.

6.3.4 Unfolding to Particle Level

Once the simulation at detector level has been ensured to describe the measurement, the simulation samples can be used to revert the effect of the detector on the cross section. The *effect of the detector* usually refers to the smearing of the p_T spectrum: indeed, resolution effects translates into migrations of jets among neighbouring p_T bins, which, in case of a significant slope, implies a smearing. One can include the dependence in other variables, such as the rapidity.

Formally, the transition from the *true* (unknown) count N_b to the measured (known) count $N_{\hat{b}}$ can be expressed in terms of a matrix that can be constructed from the simulation; in this context, the transition matrix is often called Response Matrix (RM). However, the steeply falling character of the p_T spectrum can make additionally the inversion of the RM unstable. The specific treatment of the inversion of the RM is what was already mentioned as *unfolding* earlier in the chapter.

In addition to considering the migrations of jets among different regions of the phase space due to the effect of the detector, one can consider *flavour migrations* in the unfolding. Indeed, at first sight, the inclusive b jet measurement may be seen as a repetition of the all-inclusive jet measurement with an additional procedure of discrimination on the flavour of the jets; then, the inefficiency of discrimination has to be corrected for with a bin-to-bin correction extracted from the simulation. However, as we shall explain, in this thesis, we consider the b and n counts in parallel, and include in the unfolding the migrations among flavours from detector (or tag) level to particle (or true) level.

Indeed, in general, the advantage of performing a RM-based unfolding is to reduce the dependence on the simulation, in contrast to bin-to-bin corrections, especially when the bin-to-bin corrections factors are sensibly different from unity, which can lead to significant biases in the unfolded data [12]. Usually, the unfolding is performed for spectra involving different orders of magnitude, typically the p_T spectrum. Including the rapidity is less frequent, since migrations are not expected to be large; however, it may be preferable to include it in the unfolding to reduce effects at the edges of the phase space. Regarding the flavour, the performance of the b tagging will be such that the efficiency goes down to 20–30% for $p_T \rightarrow 1\,\text{TeV}$, with a significant contamination from other flavours; obviously, a dependence on the simulation would be inevitable with a bin-to-bin correction.

Such a treatment of the flavour migrations was not used in the past CMS or ATLAS publications, where only a bin-to-bin correction was performed. A major distinction that has to be mentioned is that differences between simulation and pseudo-data in toy experiments are rarely investigated in detail; however, it is crucial that the unfolding only corrects from detector effects but does not biases the physics result to the simulation. In the present analysis, the purity of the samples will be double checked (and, eventually, corrected) and the $p_T(y)$ spectrum will be modified in order to provide an optimal description of the smearing effect of the detector.

Finally, an additional advantage of using the RM-based unfolding is that statistical uncertainties can be correctly treated. Indeed, inclusive cross sections are multi-count observables, which means that jets measured in the same event are correlated. The importance of treating correlations correctly may be somewhat limited for the measurement of the inclusive *b* jet, where rarely more than one or two *b* jets are measured; but it is absolutely crucial to treat the all-inclusive jet measurement, where up to five or six jets may be measured in the same event, and, in extenso, to treat the fraction of *b* jets in the inclusive jet production.

6.3.5 Theoretical Predictions

After the unfolding to particle level, the results can be confronted to theoretical predictions. In this analysis, we first present the comparison to the three LO predictions used to perform the analysis, i.e. PYTHIA 8, MADGRAPH and HERWIG++.

In addition, we also compare the data with NLO predictions from the POWHEG box, interfaced with PYTHIA 8 for the PS, MPI and hadronisation. For the PDF, the NNPDF 3.0 [3] set has been considered (compared to the other PDFs in Fig. 6.3). Theoretical uncertainties include variations of the PDF, variations of the scales and variations of certain tune parameters.

References

1. Ball RD et al (2013) Parton distributions with LHC data. Nucl Phys B867:244–289. https://doi.org/10.1016/j.nuclphysb.2012.10.003, arXiv: 1207.1303 [hep-ph]
2. Pumplin J et al (2002) New generation of parton distributions with uncertainties from global QCD analysis. JHEP 07:012. https://doi.org/10.1088/1126-6708/2002/07/012, arXiv:hep-ph/0201195 [hep-ph]
3. Ball RD et al (2015) Parton distributions for the LHC Run II. JHEP 04:040. https://doi.org/10.1007/JHEP04(2015)040, arXiv: 1410.8849 [hep-ph]
4. Connor PLS, Hautmann F, Jung H. TMDplotter 2.2.0. http://tmdplotter.desy.de
5. Connor PLS et al (2016) TMDlib 1.0.8 and TMDplotter 2.1.1. PoS DIS2016:039. https://doi.org/10.3204/PUBDB-2016-05953
6. Aad G et al (2011) Measurement of the inclusive and dijet cross-sections of b⁻ jets in *pp* collisions at $\sqrt{s} = 7$ TeV with the ATLAS detector. Eur Phys J C71:1846. https://doi.org/10.1140/epjc/s10052-011-1846-4, arXiv: 1109.6833 [hep-ex]
7. Chatrchyan S et al (2012) Inclusive b-jet production in *pp* collisions at $\sqrt{s} = 7$ TeV. JHEP 04:084. https://doi.org/10.1007/JHEP04(2012)084, arXiv: 1202.4617 [hep-ex]
8. Voutilainen M b tagging in CMS during LHC Run-I. Private communication
9. Chatrchyan S et al (2013) Identification of b-quark jets with the CMS experiment. JINST 8:P04013. https://doi.org/10.1088/1748-0221/8/04/P04013, arXiv: 1211.4462 [hep-ex]
10. Scodellaro L (2017) b tagging in ATLAS and CMS. In: 5th large hadron collider physics conference (LHCP 2017), Shanghai, China, 15–20 May, 2017. arXiv:1709.01290 [hep-ex]
11. Sirunyan AM et al (2017) Identification of heavy-flavour jets with the CMS detector in *pp* collisions at 13 TeV. arXiv: 1712.07158 [physics.ins-det]
12. Schmitt S (2017) Data unfolding methods in high energy physics. In: EPJ web of conferences, vol 137, p 11008. https://doi.org/10.1051/epjconf/201713711008, arXiv: 1611.01927 [physics.data-an]

Chapter 7
Analysis at Detector Level

IN THIS CHAPTER, THE ANALYSIS is described at the level of the detector, i.e. without correcting the measurement from the artefacts of the detector. The selection is discussed, as well as several calibrations and their associated systematic uncertainties. At the end of this chapter, a global picture of the content of the sample at detector level is drawn.

The double differential cross section of the all-inclusive and b inclusive productions in data and simulation is shown in Fig. 7.1, as well as the ratio of simulation to data in Fig. 7.2. On the latter plot, one can observe that the ratios for the all-inclusive (in the top) and for the b-inclusive (in the bottom) differ more and more while reaching high p_T values.

Along this chapter, we explain the different procedures applied in order to obtain the distributions in Fig. 7.1, both data and simulation: first, the trigger strategy used to record the data is described. Then, the pile-up in data is described as well as the corrections applied in the simulation. This is followed by a description of the jet selection, and a first comparison of data and MC at detector level. Finally, the effect of b tagging is described. At the end, we conclude on the differences between the ratios of the all-inclusive and b-inclusive productions.

Some additional investigations and results complete the analysis in the appendix, as well as some investigations on the Missing Transverse Energy (MET).

7.1 Trigger

The general working principles of the trigger system have already been described in Sect. 3.2.2.5. The trigger strategy of the current analysis is now described.

The inclusive jet and inclusive b jet analyses use the single-jet triggers, defined only by the minimum transverse momentum of the leading jet. QCD processes being largely dominant, all these triggers are heavily *pre-scaled*, except the one of highest

© Springer Nature Switzerland AG 2019
P. L. S. Connor, *Inclusive b Jet Production in Proton-Proton Collisions*, Springer Theses,
https://doi.org/10.1007/978-3-030-34383-5_7

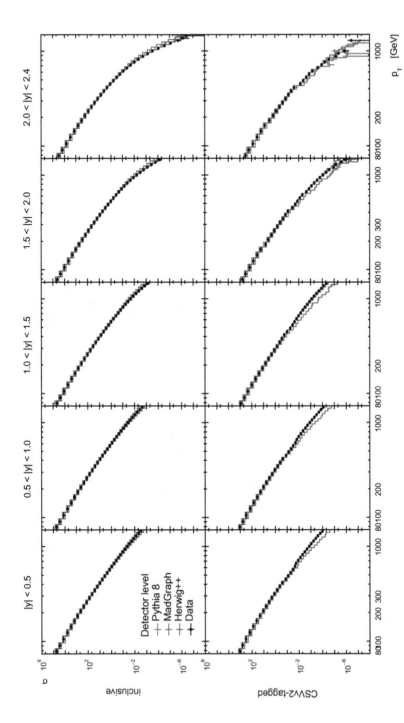

Fig. 7.1 Absolute double differential cross sections in transverse momentum and rapidity of the simulations (coloured curves) and of the data (black points with error bars). The columns correspond to the five rapidity bins; the top (bottom) row corresponds to the all-inclusive jet (inclusive $\hat{b}$ jet) selection

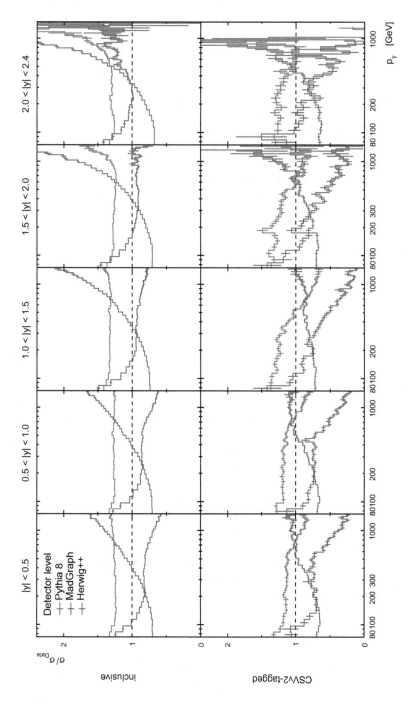

Fig. 7.2 Ratio of the double differential cross sections of the three simulation of the data. The columns correspond to the five rapidity bins; the top (bottom) row corresponds to the all-inclusive jet (inclusive b jet) selection

transverse momentum, in order to compromise between the extremely high event rate and the capability of the system to record the event. For instance, an event with a leading-jet p_T of 50 GeV will be triggered only once out of hundred thousands times, while an event with a leading-jet p_T of 800 GeV will be triggered each time.

The conditions of data taking may vary along the eras; the *trigger version* is regularly updated, corresponding to changes in the trigger pre-scales or in some correction applied on-line. Therefore, the stability of the conditions in different periods has to be checked: the four periods as defined for the Jet Energy Corrections (described later in Sect. 7.3) are BCD, EFearly, FlateG and H.

In this section, the strategy to use and combine triggers is described. First, since two triggers could be used for the same event, the strategy to combine the different triggers is given. Then the efficiencies of the different triggers are computed with a different method. Finally, the stability is checked on a run-by-run basis.

7.1.1 Trigger Strategy

The *exclusive method* is used to combine triggers such that the total cross section is the sum of the cross sections obtained from each trigger:

$$\sigma_{\text{all triggers}} = \sum_{\text{all triggers}} \sigma_{\text{trigger}} \tag{7.1}$$

In this method, each trigger is associated to a different region of the phase space. Since the triggers are defined in terms of p_T of the leading jet of the event, the phase space is simply divided as a function of the leading-jet p_T; however, all subleading jets may have lower transverse momenta.[1]

The different triggers are technically denoted as HLT_JetX_vY, where X (Y) stands for the p_T threshold in GeV (the version the trigger). Since from the trigger system a decision is needed very fast whether an event is worth being recorded, the trigger system has a very fast algorithm of reconstruction, not so precise as the PF reconstruction; therefore, p_T^{HLT}, corresponding to the HLT reconstruction, and p_T^{PF}, corresponding to the PF reconstruction need to be distinguished.

This distinction being made, the interval of p_T^{PF} (i.e. the phase space) corresponding to each trigger has to be defined. In general, the HLT algorithm being faster, it misses contributions and leads to an underestimation of the transverse momentum. This underestimation is not constant, and requires a determination of the transverse momentum from which they are fully efficient; this value is referred as *turn-on* point.

---

[1] An alternative method exists, where the phase space is divided according to the different triggers not only for the leading jet but for all jets. In this method however, the statistical accuracy is not as good as in the here adopted method, since jets would not be considered with the lowest-pre-scaled trigger available.

7.1.1.1 Determination of the Trigger Efficiency

Three methods exist to determine the trigger efficiency:

1. the *reference trigger* method,
2. the *trigger emulation* method,
3. the Tag and Probe method.

These methods will be explained in this section.

Once the efficiency has been computed, a fit of the efficiency is performed with a transformed version of the error function[2]:

$$\epsilon(p_T^{\text{PF}}) = a + 0.5 \times (1 - a) \times \left(1 + \text{erf}\left(\frac{p_T^{\text{PF}} - \mu}{\sigma}\right)\right) \tag{7.2}$$

where a, μ and σ are the fit parameters. The *turn-on* corresponds to the value of p_T^{PF} where the efficiency is 99%. A given trigger can then be used from its turn-on up to the turn-on of the next trigger.

Reference trigger method. This method is the easiest in terms of methodology. Given a trigger of a certain p_T threshold known to be fully efficient, one tests another trigger of higher p_T threshold. The efficiency is therefore obtained according to the following formula:

$$\epsilon = \frac{N(\text{test fired}|\text{ref fired})}{N(\text{ref fired})} \tag{7.3}$$

This method has the drawback of very low statistics, especially for the trigger of lowest p_T where a minimum-bias or zero-bias trigger should be used. Indeed, the reference trigger intrinsically fires less than the test trigger. Therefore, in practice, the method has not been used, but is only mentioned in order to motivate the second method.

Trigger emulation method. The second method is an improved version of the reference trigger method, with the difference that the test trigger is emulated instead of directly used. Indeed, in the reference trigger method, the statistics is limited because of the condition that both the test and the reference triggers must have fired. By reproducing the conditions in which the test trigger would have fired, one allows a larger statistics.

$$\epsilon = \frac{N(\text{test emulated}|\text{ref fired})}{N(\text{ref fired})} \tag{7.4}$$

---

[2]It is sometimes called *sigmoid function*, in reference to the S-shaped curve. However, sigmoid function refers sometimes to the *logistic function* $S(x) = \frac{1}{1+\exp(-x)}$, which is of similar shape. In comparison, the error function is defined by $\text{erf}(x) = \frac{2}{\sqrt{\pi}} \int_0^x \exp(-t^2)\, dt$. In order to avoid any confusion, the term *sigmoid function* shall be avoided.

Table 7.1 Summary of the triggers turn-on values for 99% efficiency and luminosities per trigger, rounded up to the upper edge of the corresponding p_T bin

Trigger	Turn-on	$\mathcal{L}/\text{pb}^{-1}$
HLT_PFJet_40	74	0.262
HLT_PFJet_60	84	0.711
HLT_PFJet_80	114	2.70
HLT_PFJet_140	196	23.7
HLT_PFJet_200	245	102
HLT_PFJet_260	330	580
HLT_PFJet_320	395	1730
HLT_PFJet_400	468	5070
HLT_PFJet_450	507	35200

The turn-on points are extracted with this method and are given in Table 7.1. The only remaining drawback is that the turn-on point of the trigger of lowest p_T threshold can still not be computed.[3]

Tag and Probe method. The third method allows to determine the turn-on of the trigger of lowest p_T and to cross-check the result obtained from the emulation method. The principle of the Tag and Probe method is not restricted to the determination of the trigger thresholds: it is a general method to determine the efficiency of reconstruction of a given object from situations with two such objects are expected in an event. In the present case, it consists in using events with a di-jet final state and checking when only one or both should fired the trigger.[4] First, PF jets are matched jets with HLT objects and di-jet topologies are defined as follows:

– The di-jet final state is defined such that

1. both leading jets are back-to-back: $\Delta\phi_{12} > 2.4$;
2. and all other jets have significantly lower p_T: $p_T^i < 0.3 \times \frac{p_T^1 + p_T^2}{2} \; \forall \, i > 2$.

– The matching between PF and HLT jets is defined in $\Delta R < 0.5$.

The values of parameters are empirical but safe from mismatching, in the sense that no procedure was applied in order to determine optimal values. Then the efficiency is computed as follows:

$$\epsilon = \frac{N(\text{probe}|\text{tag})}{N(\text{tag})} \tag{7.5}$$

– the *probe* jet defines whether the event has fired;
– the *tag* jet tests the trigger.

---

[3]In principle, one could use a MB trigger, but the statistics could be too low.

[4]Another typical example of Tag and Probe method is the determination of the efficiency of reconstruction for muons. In that case, candidate muons pairs with $M_{\mu\mu} \approx M_Z$ are considered: as soon as one of them has been reconstructed, it is defined as the tag; then the efficiency is determined from the rate of reconstruction of the second muon.

In practice, the efficiency depends here also on the definition of the di-jet final state; therefore the obtained value for the turn-on with the trigger emulation method should be cross-checked. The trigger efficiency for each of the four aforementioned eras is determined in bins of rapidity. The result may be found in Appendix 7.B. In general, the turn-on values do not change over time; however, they are significantly lower in the forward region, where the activity is more intense.

An example of the measurement and fit of the trigger efficiencies is shown in Fig. 7.3 for the Tag and Probe method.

7.1.1.2 Final Subdivision of the Phase Space

The choice of the turn-on points is obtained by taking the maximum given from each method, except for the first trigger (HLT_PFJet40), whose turn-on can only be computed with the Tag and Probe method. In addition, a correction is applied for the trigger inefficiency in the attempt of extending the measurement down to $p_T = 43$ GeV.

We perform a last test in order to check the turn-on points and the extension down to 43 GeV. In order to test the efficiency of the trigger, one can look for steps or irregularities in the p_T spectrum. Such an effect is expected to take place at the linear scale, and therefore cannot be seen in the logarithmic scale. A solution to find any discrepancy consists in dividing the (a priori smooth) spectrum with another smooth curve of the same order. A fit the p_T spectrum in bins of rapidity and era is performed with the following function[5]:

$$f(x) = \exp\left(a + b\log x + c\left(\log x\right)^2\right) \tag{7.6}$$

The ratio of the cross section to the fit is shown in Fig. 7.4: the fact that the ratio is not at one simply means that the function is only an approximation of the shape of the spectrum; the ratio is rather smooth at high p_T, which was expectable since conservative choices have been done, except a step that can be seen at 74 GeV, where only the Tag and Probe method could be used. The step means that the correction of the trigger efficiency is not optimal, which is most likely related to fake di-jet topologies in high pile-up conditions. Eventually, the low p_T boundary of the phase space is defined to 74 GeV, as already mentioned in Chap. 6.

The combination of the triggers to form the spectrum can be found in Appendix 7.B.

---

[5]This function is well suited for this, since it is indeed typically used for peak or step hunting, for instance in dark matter searches.

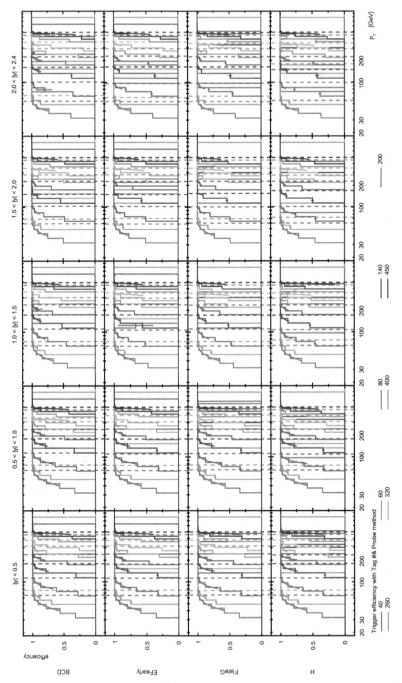

Fig. 7.3 Trigger efficiencies are obtained in bins of rapidity and fitted (continuous curves) to obtain the turn-on (vertical, dashed lines) with the Tag and Probe method. Each colour corresponds to a trigger; the row (columns) corresponds to the rapidity bins (period of data acquisition)

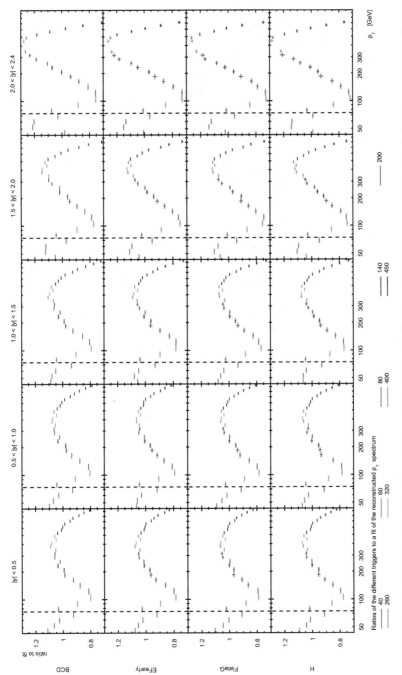

Fig. 7.4 The different contributions from the different triggers to the total cross section are divided by a smooth fit of the inclusive p_T spectrum. Each colour corresponds to a trigger; the row (columns) corresponds to the rapidity bins (period of data acquisition). The vertical, dashed, black line at 74 GeV correspond to the minimum p_T considered in the analysis

7.1.1.3   Run-by-Run Stability

A crucial test for the trigger efficiency is to check the integrated cross section for each trigger on a run-by-run basis. The cross sections are simply obtained by counting the number of jets in the events fired by the trigger and dividing it by the effective luminosity:

$$\sigma_{\text{trigger}}^{\text{run}} = \frac{N_{\text{trigger}}^{\text{run}}}{\mathcal{L}_{\text{trigger}}^{\text{run}}} \tag{7.7}$$

(One can also define the run-by-run average pre-scales, as shown in Appendix 7.B.)

The run-by-run cross section of (fraction of CSVv2-tagged jets in) the inclusive jet production is shown in Fig. 7.5 (Fig. 7.6). Apart of a few outliers, the cross section and the fraction of $\hat{b}$ jets are rather constant.

7.1.2   Effective Luminosity and Average Pre-scales

The only time dependence of the triggers is related to the conditions of data taking such as the pile-up; the pile-up decreases with time, because of the dispersion of the beam in the transverse plane. However, the hard process of interest should be of same nature at any time. The pre-scales decrease with the pile-up (already shown in Chap. 3 in Fig. 3.7). Since we are not interested in reproducing in MC the exact count as in data, we only need an average pre-scale over the whole period per trigger.

The global average pre-scale factor is given by the ratio of the effective luminosity of a trigger with the total luminosity of the sample:

$$f_{\text{trigger}}^{\text{av}} = \frac{\mathcal{L}_{\text{total}}}{\mathcal{L}_{\text{trigger}}} \tag{7.8}$$

$$= \frac{\sum_{\text{all runs}} \mathcal{L}_{\text{total}}^{\text{run}}}{\sum_{\text{all runs}} \mathcal{L}_{\text{trigger}}^{\text{run}}} \tag{7.9}$$

Eventually, the cross section of a given trigger is simply given by the following formula:

$$\sigma_{\text{trigger}} = \frac{\sum_{\text{all runs}} N_{\text{trigger}}^{\text{run}}}{\sum_{\text{all runs}} \mathcal{L}_{\text{trigger}}^{\text{run}}} \tag{7.10}$$

In fact, in addition to being simpler to apply, this procedure with average pre-scale factors has the advantage of reducing the impact of events of higher pile-up, since these usually require higher pre-scales.

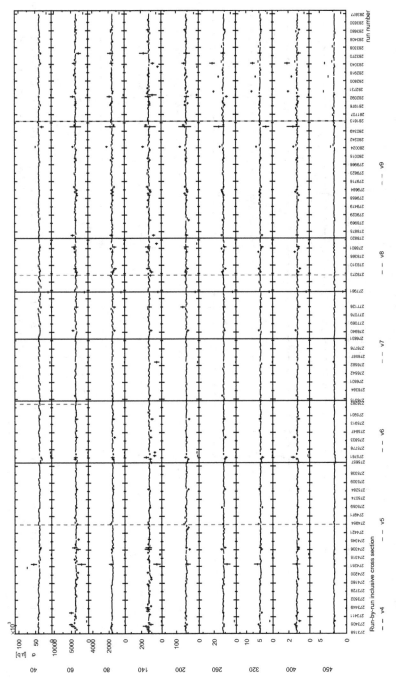

Fig. 7.5 Run-by-run inclusive jet cross section for each trigger. Each row correspond to a trigger (whose threshold is indicated on the very left). The vertical, continuous black (dashed coloured) lines correspond to a change of era (trigger version). 372 runs are considered in the analysis (only a limited number of them are explicitly labelled for readability)

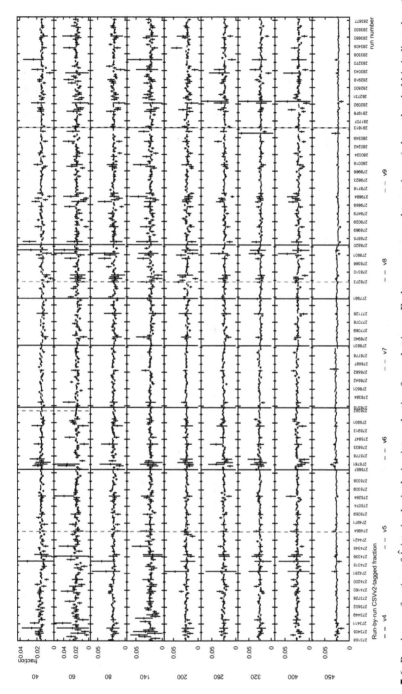

Fig. 7.6 Run-by-run fraction of $\hat{b}$ jets in the inclusive jet production for each trigger. Each row correspond to a trigger (whose threshold is indicated on the very left). The vertical, continuous black (dashed coloured) lines correspond to a change of era (trigger version). 372 runs are considered in the analysis (only a limited number of them are explicitly labelled for readability)

7.2 Pile-Up

Pile-up, already mentioned in Sect. 3.1.3, is a consequence of the increase of the delivered luminosity by LHC: on the one hand, increasing the pile-up increases the probability of seeing an interesting event; on the other hand, it increases the number of tracks in the event and may contaminate the measurement. Therefore, it is important to include a good simulation of the pile-up in the MC samples. Eventually, the effect of the pile-up on the measurement will be corrected while performing the unfolding.

In this section, general considerations about the pile-up are given. Then the two procedures applied in this analysis are described: first the reduction of overweighted pile-up events in the MC samples and secondly the reweighting procedure of the pile-up profile.

7.2.1 General Considerations on the Pile-Up

The value of the pile-up corresponds to the *number of pp interactions* per LHC bunch crossing. It has to be distinguished from the *number of vertices*:

- a *pile-up event* is a *pp* interaction,
- a *vertex* is a reconstructed point in space from which a collection of reconstructed tracks seem to come from.

The second is indeed affected by possible track inefficiencies[6] or by the reconstruction of fake vertices. In practice, for 25 interactions, one typically expects around 17 vertices [2].

From the point of view of physics, the contamination from the pile-up is two-fold:

1. additional tracks may be taken inside the jets coming from the hard process of interest;
2. additional jets may be clustered in the events and wrongly associated to the hard process of interest.

The first contamination is mitigated in the PF reconstruction by the *charged-hadron subtraction*, which removes tracks from a jet if they are associated to another vertex [3, 4]; in addition, the JEC also account for pile-up effects [5] (discussed later in Sect. 7.3). The second contamination is not crucial in the present analysis, since the measurement is done at relatively high transverse momentum; therefore, contamination from pile-up jets is not expected to be significant.

On the point of view of the treatment, we need to ensure two points:

1. the pile-up simulation has to be performed correctly;
2. the *in-time pile-up*[7] profile in simulation has to be corrected to the one in data.

---

[6]In particular, the data collected until mid-August 2016 (i.e. at the end of RunF) is affected by a dynamic inefficiency in the track reconstruction [1]. This has not been simulated in the MC samples.

[7]An *out-of-time pile-up* exists as well, coming from the overlay of successive bunch crossings.

These two points will be the objects of two reweighting procedures that will be described later in this section.

7.2.1.1   Measurement of Pile-Up in Data

The method to estimate the pile-up in data consists in exploiting the relation between cross section and luminosity $\sigma = N/\mathcal{L}$ for the MB inelastic cross section.[8] Using this method for each LS successively, one can estimate the number of interactions per bunch crossing, i.e. the pile-up. Eventually, given the instantaneous luminosity, the numbers of vertices roughly corresponds to around 70% of the number of interactions.

All CMS analyses in 2016 use the same MB inelastic cross section [6, 7]:

$$\sigma_{MB} = 69.2 \pm 3.2 \, \text{mb} \qquad (7.11)$$

7.2.1.2   Simulation of the Pile-Up

The technique consists in generating QCD events following a Poisson distribution $\mathcal{P}(\lambda)$:

$$p(k) = \frac{\lambda^k}{k!} \exp(-\lambda) \qquad (7.12)$$

However, a pure Poisson distribution can be too approximative; in practice, it is implemented as follows:

1. Several Poisson distributions with different λ parameters may be added to simulate better the data.
2. The mean of the Poisson distribution is taken a bit higher than the expected average pile-up in data; in other words, the MC samples are produced with slightly overestimated scenarios, and corrected later if need be.

In the current analysis, the PYTHIA 8 and MADGRAPH sample share the same pile-up simulation (with double Poisson distribution), while HERWIG++ has a slightly older one (with only one Poisson distribution); indeed, PYTHIA 8 and MADGRAPH account for the change of pile-up conditions after the tracker dynamics inefficiency at the end of RunF, while HERWIG++ does not [1]. In both cases, a reweighting procedure, described later in this section, is applied to improve the description of the pile-up in the simulation in agreement with the data.

---

[8] At detector level, the MB trigger is simply defined by a minimal amount of energy deposit in the HF.

7.2.2 Removal of Overweighted Events in Simulation

As already explained in Sect. 6.3.2, in order to get a high statistics for all p_T values, the PYTHIA 8 (MADGRAPH) samples are generated in slices of $\hat{p}_T$ (H_T). However, the pile-up simulation is performed in addition without slices in p_T values, therefore, high-p_T jets can be produced even for the low-p_T slices. Therefore, when rescaling to the cross section, events with high-p_T jets will largely dominate the population. To fix this intrinsic problem of the simulation, events with a "too high" reconstructed leading jet p_T are rejected; the exact procedure is described in this subsection. As a result, this procedure modifies the count N in Eq. 6.5.

Being generated in slices of H_T rather than in slices of $\hat{p}_T$, the slices of MADGRAPH are wider with respect to the p_T spectrum. Nevertheless, a similar cut-off is applied on the leading generated jet (the $\hat{p}_T$ can be defined but does not allow to define exclusive samples, and is therefore not useful in the MADGRAPH samples).

The same issue of outliers in the simulation happens in HERWIG++. However, even though the sample is not generated in slices, they are more problematic, as the events are generated uniformly with respect to $\hat{p}_T$, and as the statistics is much smaller.

The cut-off is defined as follows: if the transverse momentum of the leading reconstructed jet is a few times higher than the transverse momentum of the leading generated jet (or than the scale of the ME), then the reconstructed jet is considered as a *bad jet* and has to be removed from the simulation:

$$X \times p_T^{\text{gen,lead}} < p_T^{\text{det,lead}} \tag{7.13}$$

$$X \times \hat{p}_T < p_T^{\text{det,lead}} \tag{7.14}$$

Unfortunately, with this cut-off purely based on the transverse momentum, it is not possible to remove these bad jets without removing as well *good jets*, i.e. without changing the shape of the spectrum. Therefore, an additional uncertainty is estimated by varying the value of the cut-off; in practice, it seems that $X = 3.5 \pm 1.0$ allows to remove the bad events with limited impact on the simulation.

It was also checked whether only the jet outliers should be removed or the full event containing the jet outlier. It turned out that this choice has a much lower effect than the choice of the numerical value of the cut-off.

In Fig. 7.7a, the effect of the procedure is shown on the absolute cross section from PYTHIA 8; from the colours, it can be seen that the bad jets are removed from the spectrum. The effect of the cut-off together with the associated uncertainties can be seen on Figs. 7.7b for the three MC samples. PYTHIA 8 and MADGRAPH are affected only at small p_T with around one percent; however, HERWIG++ is extremely contaminated.

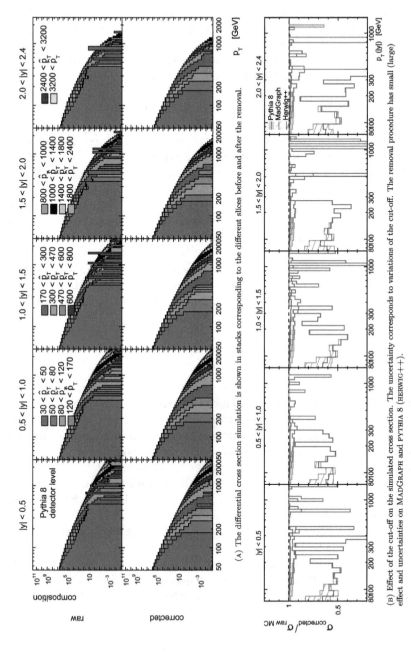

(A) The differential cross section simulation is shown in stacks corresponding to the different slices before and after the removal.

(B) Effect of the removal of the overweighted pile-up jet outliers and bins of rapidity. The different columns correspond to the rapidity bins effect and uncertainties on MadGraph and Pythia 8 (Herwig++).

Fig. 7.7 Effect of the removal of the overweighted pile-up jet outliers and bins of rapidity. The different columns correspond to the rapidity bins

7.2.3 Procedure of Pile-Up Reweighting

The procedure of reweighting to the observed pile-up is applied to reproduce the pile-up conditions in the MC samples. The distribution on which this procedure is applied is the so-called *pile-up profile*, which describes the pile-up as a function of the average number of bunch crossings. At CMS, the pile-up profile is computed with an independent framework from the one used to perform the analysis [7]; however, it was taken care to extract it from the same data sample (i.e. from the full 2016 period of data taking).

Three quantities are shown in Fig. 7.8a before and after the reweighting procedure:

1. The pile-up profile is shown in the first column; the agreement after calibration is perfect as expected.
2. The number of (good) vertices is shown in the second and third columns; the agreement after calibration cannot be perfect as expected (since for instance, as mentioned earlier in the section, the tracker dynamics inefficiency is not simulated), but is improved.
3. Finally, the additional soft activity due to the pile-up is well described by the ρ variable [8]:

$$\rho = \text{median} \left[\left\{ \frac{p_{tj}}{A_j} \right\}_{j=1,\ldots,N_{\text{jets}}} \right] \tag{7.15}$$

where A corresponds to the *jet area* [9]. Since it is a median, ρ is not sensitive to the hard activity, and estimates the UE, the electronics noise, and the pile-up. Though not perfect, the agreement is improved, and the uncertainties reduce the difference between data and simulation.

The uncertainty band corresponds to the uncertainty on the MB measurement (Eq. 7.11).

The pile-up profile is then reweighted to correct the simulation to the data. The effect on the double differential cross section can be seen in Fig. 7.8b. Since the simulation of the pile-up in PYTHIA 8 and MADGRAPH (HERWIG++) is close to (different from) the real pile-up in 2016, the reweighting has a small (large) effect on the cross section; the width of the uncertainty band is related to the statistics of the sample, therefore larger for HERWIG++.

7.3 Jets

Jets were introduced in Sect. 2.3.2.2. The interest in jets relies in that they can be defined at both particle and detector level. Elements of reconstructions were given in Sect. 3.2.2.2.

At CMS, the performance of the jet reconstruction is studied centrally [10, 11]. In this section, we discuss the calibration of jets both in data and simulation: in data, the

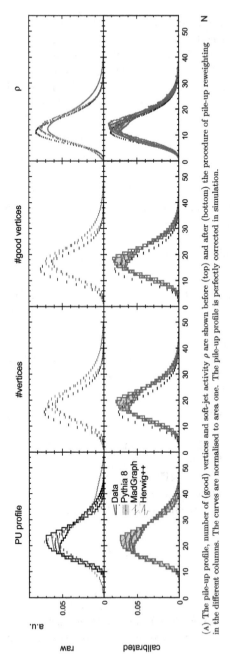

(A) The pile-up profile, number of (good) vertices and soft-jet activity ρ are shown before (top) and after (bottom) the procedure of pile-up reweighting in the different columns. The curves are normalised to area one. The pile-up profile is perfectly corrected in simulation.

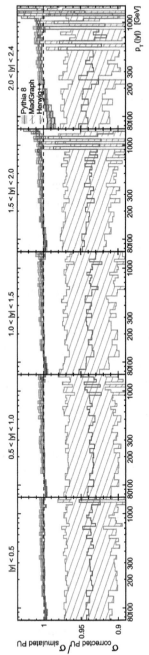

(B) Effect of pile-up reweighting with associated uncertainties on the spectrum. The reweighting procedure has small (large) effect and uncertainties on MadGraph and Pythia 8 (Herwig++). The different columns correspond to the rapidity bins.

Fig. 7.8 Effect of the reweighting of the pile-up profile of the simulation to the pile-up profile in the data. The uncertainty bands correspond to the uncertainty on the measurement of the MB correction (Eq. 7.11) and is red/shaded for Pythia 8, blue/hashed for MadGraph and purple/hashed for Herwig++

dependence on time needs to be compensated; in simulation, it needs to describe the response of the detector in data. Then, we discuss the resolution on the jet kinematics in data and simulation.

In Appendix 7.E, we check the jet constituents by investigating multiplicity and the energy fraction of the different stable particles entering into the jet composition.

7.3.1 Jet Energy Scale Correction

The purpose of Jet Energy Corrections (JECs) is to correct the measured to the true energy of jet in the form of a global multiplicative factor to the four-momentum:

$$p_\mu^{\text{true}} = C_{\text{JEC}} \times p_\mu^{\text{raw}} \tag{7.16}$$

This correction factor can be further divided into several components, applied in a chain.

$$C_{\text{JEC}} = C_{\text{offset}}(p_T^{\text{raw}}) \times C_{\text{MC}}(p_T', \eta) \times C_{\text{relative}}(\eta) \times C_{\text{absolute}}(p_T'') \tag{7.17}$$

The corrections are computed at three different *levels* in sequence, as illustrated by a diagram in Fig. 7.9, in the following order:

1. The *offset correction* C_{offset} removes everything that is not related to the pp collision of the hard interaction, e.g. pile-up and electronic noise ($p_T^{\text{raw}} \rightarrow p_T'$).
2. The *MC calibration* C_{MC} corrects for the main non-uniformities in pseudorapidity and non-linearities in transverse momentum ($p_T' \rightarrow p_T''$); typically, the calorimeters have a non-linear response.
3. The *residual corrections* $C_{\text{relative}} \times C_{\text{absolute}}$ accounts for finer corrections between data and MC ($p_T'' \rightarrow p_T^{\text{true}}$):

   - relative energy scales are corrected by investigating dijet topologies, where the same energy is expected from both jets in opposite direction.
   - absolute energy scales are corrected by investigating $Z +$ jets topologies, where the energy is measured accurately from the Z decay into electrons in the ECAL or into muons in the tracking system; complementarily, multijet events are also used.

An additional correction for HF jets can also be included, to account for differences with light jets, but the effect is mostly relevant for low p_T and is not considered here.

At CMS, the JEC are provided centrally [5, 11], together with an estimation of the associated uncertainties. The current JEC are computed with PYTHIA 8, therefore suited for the PYTHIA 8 and MADGRAPH samples, but not totally adequate for the HERWIG++ sample.

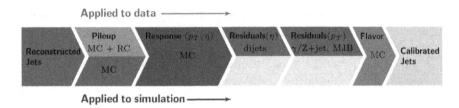

Fig. 7.9 Diagrammatic description of the application of JECs. More corrections are applied to data than to MC. Additional flavour corrections are not considered in this analysis. RC stands for *Residual Corrections*. Taken from [11]

Each level contributes to the uncertainties:

1. The pile-up offset is mainly important at low p_T, and its uncertainty is below 0.1% from $p_T \sim 100\,\text{GeV}$.
2. The time stability matters especially at high p_T, where it reaches 0.5% of the uncertainty.
3. The absolute (relate) scale contributes for around slightly less than 1% (slightly more than 0.1%) of the uncertainty.

Here we only show the global uncertainties due to JEC on the $p_T(y)$ spectrum in Fig. 7.10a.

7.3.2 Jet Energy Resolution

The choice of the binning scheme is related to the Jet Energy Resolutions (JERs); in QCD measurements at CMS, the binning is standard for all jet analyses in order to ease the comparison of measurements. The resolution on the transverse momentum is usually finer in MC than in data; therefore the transverse momentum has to be smeared, which is crucial for the unfolding. In the current subsection, we explain the procedure of smearing of the spectrum of transverse momentum in MC and check that the standard binning is reasonable. The binning is given in Appendix 7.C.

7.3.2.1 JER Smearing

In MC samples, given a jet generated (reconstructed) with a transverse momentum p_T^{gen} (p_T^{rec}), we define the *resolution* as follows:

$$\Delta = \frac{p_T^{\text{rec}} - p_T^{\text{gen}}}{p_T^{\text{gen}}} \qquad (7.18)$$

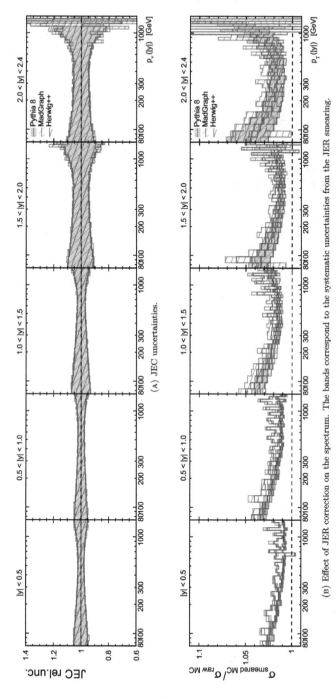

(A) JEC uncertainties.

(B) Effect of JEC and effect of smearing of the transverse momentum on the spectrum for all three simulations. The bands correspond to the systematic uncertainties from the JER smearing.

Fig. 7.10 Uncertainties of JEC and effect of smearing of the transverse momentum on the spectrum for all three simulations. The columns correspond to the rapidity bins; the bands correspond to the systematic uncertainties

For a given p_T^{gen}, it is a Gaussian-like curve with a core and tail, but with two characteristics:

1. it is slightly asymmetric, since it is more probable for a jet to be reconstructed at lower values, when components (tracks) are missed by the reconstruction;
2. the left tail is more important because of various reconstruction effects (e.g situations where a jet is reconstructed into two jets, or situations where a pile-up jet is considered by mistake).

Analytically, the curve is usually fitted with a simple Gaussian curve (or with a Crystal-Ball curve, the latter being a modified version of the Gaussian curve to take into account the behaviour of the left tail). Therefore, the choice of this function may be discussed, since is does not describe the deviation to the Gaussian curve from the top; however in practice, a simple Gaussian fit appears to be enough.

Sometimes, the term *resolution* refers specifically to the width σ_{JER} of this Gaussian-like curve.

The resolution in data is measured and released centrally at CMS. Given these resolutions, one has to make sure that they are similar in data and MC by applying smearing on the spectrum of transverse momentum.

In principle, given the resolution Δ, the reconstructed and generated transverse momenta is related by the following formula (which is a rewritten version of Eq. 7.18):

$$p_T^{\text{rec}} = p_T^{\text{gen}} \times (1 + \Delta_{\text{MC}}) \qquad (7.19)$$

Two methods exist to correct the resolution [12]:

Scaling method   This method assumes that the following matching can be done:
- $\delta R < R_{\text{cone}}/2$ where the R_{cone} is the cone size radius of the jet clustering algorithm (here $R_{\text{cone}} = 0.4$), and $\delta R = \sqrt{\delta y + \delta \phi}$ is the angular separation;
- $|\Delta_{\text{MC}}| < 3\sigma_{\text{JER}}$ where σ_{JER} is the measured resolution in data.

Then the resolution obtained from the MC value of p_T^{rec} has to be corrected with Scale Factors (SFs):

$$\Delta_{\text{data}} = \text{SF} \times \Delta_{\text{MC}} \qquad (7.20)$$

Given this correction to the resolution, the value of p_T^{rec} can be corrected in turn:

$$p_T^{\text{rec}} = p_T^{\text{gen}} \times (1 + \Delta_{\text{MC}}) \qquad (7.21)$$
$$\longmapsto \ p_T^{\text{rec}} = p_T^{\text{gen}} \times (1 + \text{SF} \times \Delta_{\text{MC}}) \qquad (7.22)$$

These *smearing scale factors* are extracted from the measurement of the resolution in data, and are provided centrally as well.[9]

Stochastic method   This alternative method is intended to be used in the case no matching can be done; in this case, one resorts to random numbers. One picks

---

[9]In practice, it may sometimes be easier to apply $p_T^{\text{rec}} \longrightarrow p_T^{\text{rec}} \times (1 + (\text{SF} - 1) \times (p_T^{\text{rec}} - p_T^{\text{gen}})/p_T^{\text{rec}})$. The two formulations are entirely equivalent — one just has to play with the definition of the resolution.

a number from the data resolution according to a centred Gaussian distribution with width equal to the resolution σ_{JER}, i.e. one picks $\Delta \sim \mathcal{N}(0, \sigma_{JER})$; then the transverse momentum is smeared accordingly:

$$p_T^{rec} \longmapsto p_T^{rec} \times \left(1 + \Delta \times \sqrt{\max\left(SF^2 - 1, 0\right)}\right) \qquad (7.23)$$

In practice, a so-called *hybrid method* is applied, according to whether the matching may be performed or not.

The effect on the spectrum can be seen on Fig. 7.10b: all simulations give similar results; the resolution has much larger uncertainties in the forward region than in the central, which is related to the contamination from the pile-up in the forward region; however, the smearing is more important in the central region than in the forward region.

7.3.2.2 Differential Resolution and Binning

After the JER smearing, the resolution can be checked. The differential resolution is shown in Fig. 7.11a for PYTHIA 8 in the five rapidity bins.

The profile of the resolution is extracted from the differential resolutions in order to check the binning, as shown in Fig. 7.11b. For each p_T bin, a Gaussian fit is performed, and the mean $\langle \Delta \rangle$ (top row) and width $\sigma [\Delta]$ (bottom row) are extracted. The resolution defined by the bin width is shown with a black curve and correspond to the following:

$$\frac{\delta p_T}{p_T} \approx \frac{p_T^{up} - p_T^{down}}{p_T^{center}} \qquad (7.24)$$

where up, down and center are defined for each bin respectively.

In average, the jets are reconstructed in the same bin and the jet migrations follow the bin widths; the width of the profile is a multiple of the bin width, except for low values where the bin width is less regular. The systematic shift of HERWIG++ with respect to PYTHIA 8 and MADGRAPH can be explained by the fact that there is no proper corrected to HERWIG++ (i.e. the SFs are determined only for PYTHIA 8).

It is also shown in Appendix 7.C for the systematic uncertainties of JER and JEC and for b and $\hat{b}$ spectra.

7.4 Tagging

The principles of b tagging were already introduced in Sect. 5.3.1.2. We saw that different properties of b jets can be exploited to identify (or *tag*) them: the presence of a SV, the impact parameters of the tracks or the presence of a non-isolated lepton in the jet.

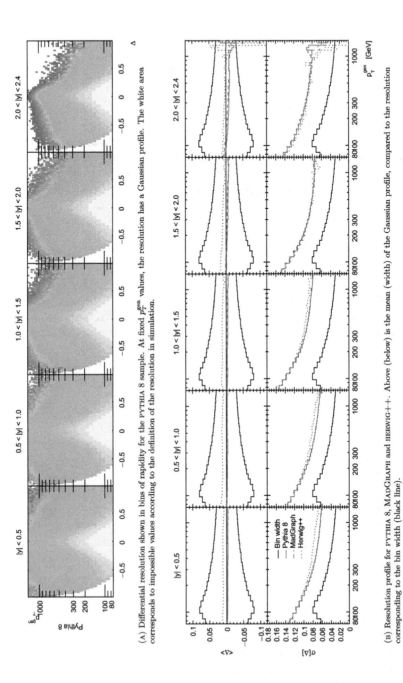

(A) Differential resolution shown in bins of rapidity for the PYTHIA 8 sample. At fixed p_T^{gen} values, the resolution has a Gaussian profile. The white area corresponds to impossible values according to the definition of the resolution in simulation.

(B) Resolution profile for PYTHIA 8, MADGRAPH and HERWIG++. Above (below) is the mean (width) of the Gaussian profile, compared to the resolution corresponding to the bin width (black line).

Fig. 7.11 Differential resolution and resolution profile, shown in bins of rapidity (columns) as a function of the transverse momentum of the generated jets

In this section, we describe the taggers at CMS in more detail and explain the calibration of the b tagging in the simulation.

7.4.1 Taggers at CMS

At CMS, different taggers are defined centrally [13–15]; in this section, we present three of them: Combined-Secondary-Vertex (CSVv2), Jet Probability (JP) and combined-Multi-Variate-Analysis (cMVAv2).

7.4.1.1 CSVv2

The CSVv2 tagger primarily exploits the presence of a SV with an invariant mass $M_{SV} < 6.5$ GeV i.e. around 1.5 GeV above the mass of the B hadron (the invariant mass is computed from the tracks associated with the SV). To the difference of the PV that is reconstructed with the Adaptive Vertex Reconstruction (AVR) fitter (see Sect. 3.2.2.1), the SV is by default based on the IVF: indeed, the AVR fitter is only based on tracks already clustered in jets, while the IVF takes all tracks into account. Pairs of tracks compatible with a long-lived K_S^0 are rejected. Finally, the flight direction of the SV must satisfy an angular separation $\Delta R < 0.4$ with the axis of the jet.

In its first version, during Run-I, the CSV tagger was also *combined* (hence its name) with additional variables, in order to increase the performance; for instance, it could be combined with track-based information in case no SV could be reconstructed, in a similar way as the JP tagger (described in the next paragraph). In its second version, intended for Run-II, many additional variables have been added in order to increase the power of discrimination: for instance a correction to the SV mass is considered, as well as various variables describing the kinematics of the tracks belonging to the SV and its geometry. All the variables are combined into a neural network.

In the present version, in this work, the CSVv2 tagger is trained in 19 (p_T, $|y|$) bins, which are given in Table 7.2. Its shape is sketched in Fig. 7.12a: it can be seen that it takes the shape of a valley, where light jets (b jets) are concentrated in the left (right). It is the main tagger for this analysis.

7.4.1.2 Jet Probability

The SV is sometimes too close to the PV to be identified with a SV tagger. Alternatively, the JP tagger is based on the impact parameter of the displaced tracks with respect to the PV. This technique is inherited from LEP [16, 17], with a slight difference in the definition in order to be suitable for pp collisions.

Table 7.2 Binning of the training of the CSVv2 tagger

p_T/GeV	$|\eta|$
15−40	0−1.2−2.1−2.4
40−60	0−1.2−2.1−2.4
60−90	0−1.2−2.1−2.4
90−150	0−1.2−2.1−2.4
150−400	0−1.2−2.1−2.4
400−600	0−1.2−2.4
600−∞	0−1.2−2.4

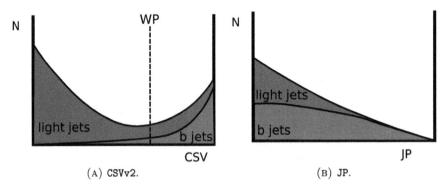

(A) CSVv2.   (B) JP.

Fig. 7.12 Sketches of discriminant variables. The red (blue) corresponds to bottom (light) jets. The discriminant takes different shapes according to the flavour. The Working Point (WP) is an arbitrary point separating flavours. Charm jets are neglected in this picture

The construction of the JP tagger can be summarised as follows: given a jet, a probability for each track of the PV to belong to the jet is first defined from the resolution; indeed, tracks coming from a SV will rather populate the tail of the resolution function. The track probabilities are then combined into a probability for a jet; the combination is defined so as to be sensitive to the presence of tracks with low probability to below to the PV, i.e. to be sensitive to the presence of B mesons. (Despite its name, the JP tagger itself is rather the (negative) logarithm of the jet probability; therefore its value can be greater than one.)

The distribution takes the shape of a descending slope, where light jets mostly peak to 0 while b jets are spreader, as illustrated in Fig. 7.12b.

7.4.1.3   cMVAv2

The cMVAv2 tagger combines several taggers with MVA techniques, in order to exploit the advantages of all the techniques. It combines two versions of the CSVv2 tagger, two versions of the JP tagger and two additional soft-lepton taggers; the

Fig. 7.13 "The correlation between the different input variables for the cMVAv2 tagger for *b* jets in *t t̄* events. (...)" [15] JBP is a variant of JP; AVR and IVF are two vertex fitters; SM (SE) stands for *Soft Muon (Soft Electron)*

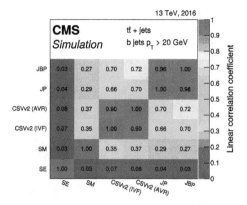

correlations among the taggers are shown in Fig. 7.13. In this analysis, it will be considered as a cross-check.

7.4.2 Performance of the Taggers

As already mentioned in Sect. 6.3.1, different levels of discrimination can be applied, corresponding to different compromises between the efficiency to tag a *b* jet and the rate of (mis)tagging a light or *c* jet; three standard *Working Points (WPs)* are usually defined for jets with $p_T > 30$ GeV as follows:

tight (T) for 0.1% of misidentified light jets,
medium (M) for 1% of misidentified light jets,
loose (L) for 10% of misidentified light jets.

The misidentification rates increase significantly at higher p_T, as will be shown in this section. The reason for defining these parameters is to compute standard calibration within the CMS collaboration. In this analysis, since the aim is to measure *b* jets with high p_T, the tight WP shall be considered; the two other WPs will be used for cross checks.

In this subsection, we explain the calibration of the MC samples, then we show the performance of tagging techniques in data.

7.4.2.1 Calibration

The calibration consists in correcting the MC efficiency and misidentification rates to the data [18]; the effect is illustrated in Fig. 7.14. The corrected quantity is the count N in Eq. 6.5; however, it only rearranges the contributions from the different flavours but does not change the all-inclusive double differential cross section, i.e., it changes $N_{\hat{b}}$ and $N_{\hat{n}}$ but not their sum.

Fig. 7.14 Illustration of the procedure of calibration on a discriminant variable. The red (blue) corresponds to bottom (light) jets. The calibration rescales the different contributions to the total cross section in bins of transverse momentum, correcting the efficiency of the simulation to the efficiency of the data; the total cross section does not change. Charm jets are neglected in this picture

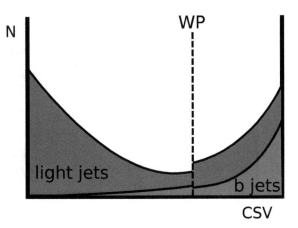

Procedure. The calibration is described only as a function of p_T and of the true flavour; in principle, it should be also a function of η, but this was not done at CMS in 2016. It is formulated in terms of SFs, given for each flavour; *b-tag SFs* denote the SFs for tagging (true) b jets, while *mistag SFs* denote the SFs for tagging light and c jets. A reweighting factor w_{entry} is computed for each entry from the SFs and from the efficiencies and misidentification rates:

$$
w_{\text{entry}}(p_T, \eta, f) = \frac{\prod\limits_{i \in \text{tagged}} SF_i(p_T, f)\epsilon_i(p_T, \eta, f) \prod\limits_{j \in \text{non-tagged}} (1 - SF_j(p_T, f)\epsilon_j(p_T, \eta, f))}{\prod\limits_{i \in \text{tagged}} \epsilon_i(p_T, \eta, f) \prod\limits_{j \in \text{non-tagged}} (1 - \epsilon_j(p_T, \eta, f))}
$$

(7.25)

where f stands for the flavour. The inclusive b jet analysis is a multi-count observable, i.e. each jet corresponds to one single entry; therefore, the computation of the weight simplifies to the following[10]:

– if the jet is tagged: $w_{\text{jet}}(p_T, \eta, f) = SF(p_T, f)$
– else: $w_{\text{jet}}(p_T, \eta, f) = (1 - SF(p_T, f)\epsilon(p_T, \eta, f))/(1 - \epsilon(p_T, \eta, f))$

(In practice, the singularity at 1 is always avoided, since the efficiency hardly reaches 0.6.) The point of Eq. 7.25 is simply that the inclusive jet spectrum remains unchanged by the procedure: the flavours are only rebalanced inside of each $(p_T, |y|)$ bin.

Scale factors. The SFs correspond to ratios of efficiencies in MC and data, and are obtained from specific processes where the efficiency can be measured in data; the SFs are shown in Fig. 7.15 for the tight WP (see Appendix 7.D for the other WPs). The correction applied to b jets is close to one, as well as for c jets since they are applied the same SFs with larger uncertainties; however, the correction applied to

---

[10]For instance, in the case of the measurement of the mass of the $b\bar{b}$ pair, two b jets would enter the same bin; therefore the computation of the weight would be more sophisticated.

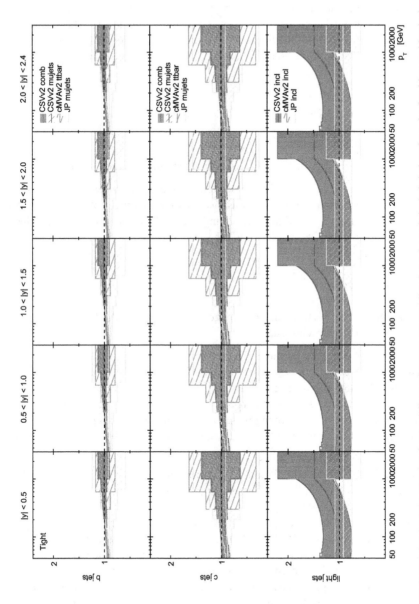

Fig. 7.15 Scale factors for the tight selection for the different flavours. No rapidity dependence is included, therefore, the five bins are identical

light jets gets larger and larger at higher p_T, corresponding to boosted topologies where the discrimination is more difficult.

Uncertainties. The procedure of calibration comes with uncertainties related to the SFs. The determination of the SFs is performed by performing template fits[11]; the templates are obtained from the simulation, and all the uncertainties on the simulation that may affect the shape of the templates are propagated to determine uncertainties on the SFs. Variations of the shape of the templates are the following:

– scale variation,
– background subtractions,
– JEC and JER variations,
– pile-up reweighting,
– fragmentation and Gluon Splitting (GSP).

The two last ones are the largest sources of uncertainties in the b calibration.

Application. The application of the procedure on the $p_T(y)$ spectrum can be seen in Fig. 7.16 on the inclusive b jet; it was also checked that the global effect of this reweighting procedure indeed does not change the inclusive jet cross section, since it only corrects the tagging efficiency in the simulation to the one in data.

7.4.2.2 Performance

In order to assess the performance of a tagger and to compare the MC samples, one needs to define the following quantities:

fraction   fraction tagged jets among all jets, i.e. $f = N(\text{tagged})/N(\text{all})$
fraction ratio   ratio of the respective fractions in MC and data, i.e. $f_{\text{MC}}/f_{\text{data}}$
efficiency   fraction of tagged jets among the true jets, i.e. $N(\text{tagged}|\text{true})$
mistag   fraction of tagged jets among the non-true jets, i.e. $N(\text{tagged}|\text{non-true})$
purity   fraction of true jets among the tagged jets, i.e. $N(\text{true}|\text{tagged})$
contamination   fraction of non-true jets among the tagged jets, i.e. $N(\text{non-true}|$
        tagged)

These quantities will be essential for the discussion; in particular, it is crucial to understand the fraction. Indeed, it allows to estimate whether the calibration of b tagging in the simulation is correct.

The performance is shown for the PYTHIA 8 and MADGRAPH samples after tagging with CSVv2 at the tight WP (CSVv2T) in Fig. 7.17, including uncertainties from SFs; the other taggers and WP can be seen in Appendix 7.D. One can observe a disagreement especially at high p_T, particularly marked in the region $1.0 < |y| < 1.5$. In the different appendices of this chapter are extensive studies conducted in order to investigate possible reasons. Finally, the disagreement will be treated in Chap. 8.

---

[11] A similar procedure is performed in Chap. 8.

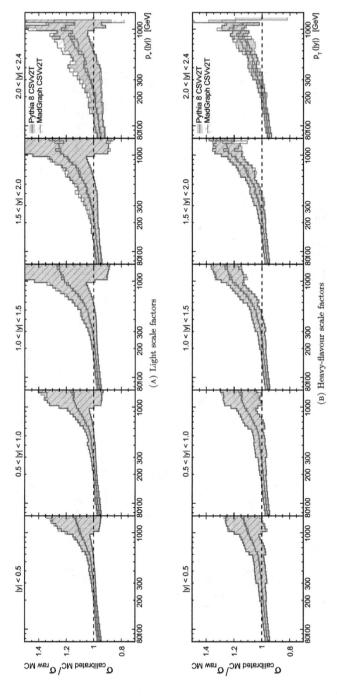

Fig. 7.16 Effect of the b calibration on the spectrum compared for PYTHIA 8 and MADGRAPH. Charm and bottom jets share the same scale factors, as they have very similar properties (charm have doubled uncertainties). The five columns correspond to the rapidity bins

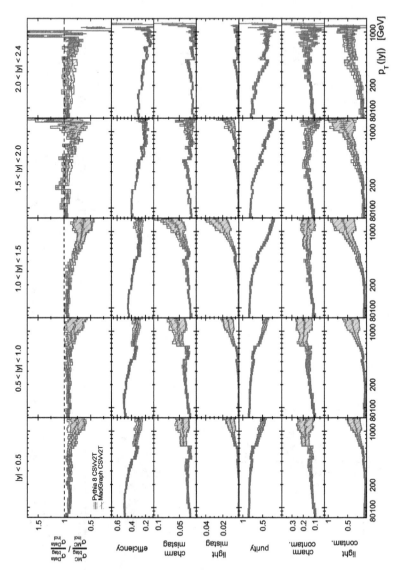

Fig. 7.17 Performance of CSVv2 in simulation after calibration with the tight selection. Calibration is only available for CUETP8M1. A significant disagreement from one in the fraction ratio can be seen for both simulations

A small discrepancy is also seen in all rapidity bins for $p_T = 400-600$ GeV; it occurs in all quantities except the fraction ratio where it cancels out; it corresponds most likely to a problem in the training of the tagger[12]; the discrepancy will be solved in the unfolding procedure.

Note that as the calibration is provided assuming PYTHIA 8 hadronisation, an additional calibration would be needed to include HERWIG++. This has not been done here; however, the effect of the calibration not being too large (though still present), including HERWIG++ in the investigations help assessing the agreement, since it is completely independent from PYTHIA 8 and MADGRAPH; this is shown in Appendix 7.D.

Conclusions

In this chapter, we have investigated in detail the samples of data and simulation. The trigger efficiency has defined the low-p_T boundary of the phase space; the pile-up in the simulation has been corrected to the real pile-up profile; the response of the detector in the jet reconstruction has been described and corrected; finally, the tagging has been investigated. Uncertainties have been associated to all corrections in the simulation.

The difference in the simulation-to-data ratios (Fig. 7.2) for the inclusive jet and inclusive b jet production is confirmed in all situations, regardless of the tagger or of the WP, and does not seem to be related to the sample, since the HERWIG++ sample has confirmed tendencies from PYTHIA 8 and MADGRAPH.

7.A   Missing Transverse Energy

The *missing transverse momentum* corresponds to the momentum vector imbalance in the perpendicular plane (xy) to the beam axis (z):

$$\mathbf{p}_T^{\text{miss}} = -\sum_{\text{jets}} \mathbf{p}_T^{\text{jet}} \tag{7.26}$$

The Missing Transverse Energy (MET) is the magnitude of the missing transverse momentum:

$$E_T^{\text{miss}} = |\mathbf{p}_T^{\text{miss}}| \tag{7.27}$$

---

[12]Unfortunately, as it can be seen in Appendix 7.D, it reflects also in the performance of the CMVAv2 tagger. The CMVAv2 can therefore not be used as a substitute.

Some physics processes or detector effects may cause it to be significant. In this section, we check these effects in the inclusive jet and inclusive *b*-tagged jet reconstructed spectra, first at generator level, then at detector level.

7.A.1   Detector Studies

General effects have already been studied in the inclusive jet measurements at CMS at 7 and 8 TeV [19, 20]. In the CMS publication at 8 TeV [21], a cut-off on the fraction of MET is applied in order to reduce the contribution from event suffering from significant detector effects with the least effect on physics processes. However, such a cut-off is not applied in the inclusive-jet analysis at 13 TeV with 2015 data [22], nor in the inclusive-*b*-jet analysis at 7 TeV [23]. The cut-off is very much dependent on the condition of data taking and needs to be checked on a case-by-base basis.

7.A.1.1   MET Variables

We first check three variables:

– the MET itself,
– the fraction of MET with respect to the total transverse energy;
– and the azimuthal angle of the MET.

The ratio of MC samples with data is shown in Fig. 7.18. Three series of curves, corresponding to the three rows, are investigated:

1. the three usual MC samples,
2. the inclusive and tagged samples in PYTHIA 8,
3. and the inclusive sample in PYTHIA 8 with different cut-offs on the fraction of MET (0.2, 0.3, 0.4).

In the first row, HERWIG++ shows a slightly different behaviour while PYTHIA 8 and MADGRAPH are very similar, which is most likely related to the respective simulations of the detector (as we said in Sect. 7.2, the simulation of the CMS detector is older for HERWIG++). The dependence on the azimuthal angle of the MET shows a phase, which is likely due to the simulation of the position of the interaction point; since the final measurement does not depend on the azimuthal angle, this phase is not relevant. The second row shows that the tagging behaves similarly for data and MC, therefore the same agreement is seen. Finally, the different cut-offs act the same way for data and MC. The conclusion is that in general, the MET is well simulated in MC, since the agreement with data does not change.

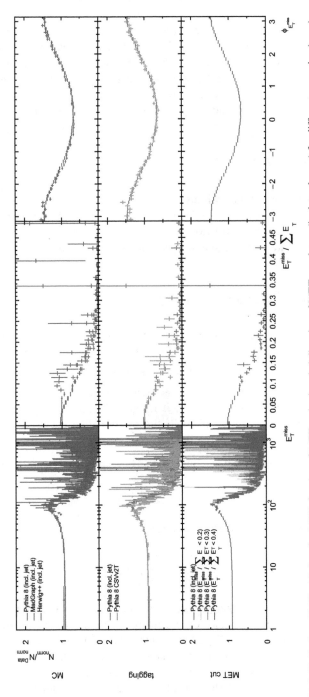

Fig. 7.18 Ratios of MC with data for MET, fraction of MET and azimuthal direction of MET are shown (in the columns) for different scenarios (rows)

7.A.1.2   Spectrum of Reconstructed Transverse Momentum

The effect of the cut-off on the fraction of MET for the inclusive jet and inclusive
b-tagged jet sample is shown in Figs. 7.19 and 7.20. In general, an effect starts
being visible, though very small, above 1.5 TeV. Since the final measurement will be
limited to the region where the calibration of b-tagging discrimination is available, it
turns out that the cut-off does not seem relevant for this analysis. However, the effect
of the cut-off on different physics processes still needs to be checked.

7.A.2   Generator Studies

The effect of the standard cut-off is also investigated at generator level in MC studies
with PYTHIA 8; it is shown in Fig. 7.21, but the same conclusions may be drawn from
other bins, including the contribution of other SM processes: $t\bar{t}$, $W +$ jet or $Z +$ jet
a.k.a. DY. Effects are only found at low transverse momentum.

7.B   Details of Trigger Efficiency

Tables for turn-on values. In Sect. 7.1.1.1, while presenting the determination of
the trigger efficiencies, we presented the emulation and the Tag and Probe methods;
they are given in Table 7.3 for each trigger, per era and per rapidity bin.

Representation of the subdivision of the phase space. In Sect. 7.1.1.2, the final
choice of the turn-on values was described. In Fig. 7.22, one can visualise the subdivision of the phase space. In complement, the number of events per trigger is given
in Table 7.4 per era and per rapidity bin.

Run-by-run average pre-scales. One can define the pre-scales as follows:

$$f_{trigger}^{run} = \mathcal{L}_{total}^{run} / \mathcal{L}_{trigger}^{run} \qquad (7.28)$$

Then, Eq. 7.7 can also be written as follows:

$$\sigma_{trigger}^{run} = \frac{f_{trigger}^{run} N_{trigger}^{run}}{\mathcal{L}_{total}^{run}} \qquad (7.29)$$

The run-by-run averaged trigger pre-scales in Fig. 7.23; as it can be seen, the variations are significant. (Similarly, as it was already mentioned in Chap. 3, Fig. 3.7, the
luminosity can become smaller by a factor of two during a single run, and pre-scales
can be adapted on-line to compensate.) Higher instantaneous luminosity implies
higher pile-up conditions, and therefore higher pre-scales. The figure shows that,
along the year of 2016, the LHC has achieved better and better performances in
terms of luminosity.

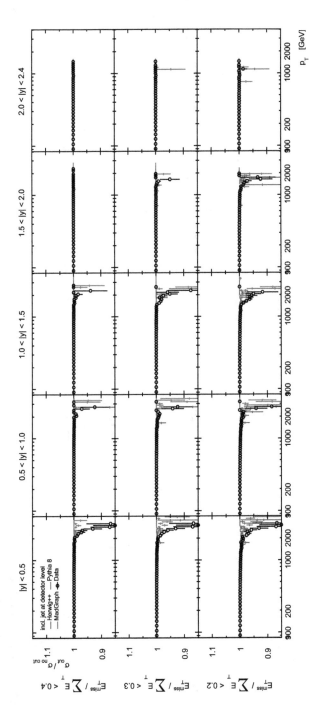

Fig. 7.19 The ratios of different samples after and before applying the different values of cut-off on the MET fraction are shown for MC and data. For the inclusive jet selection, an effect may be seen starting from 1.5 TeV

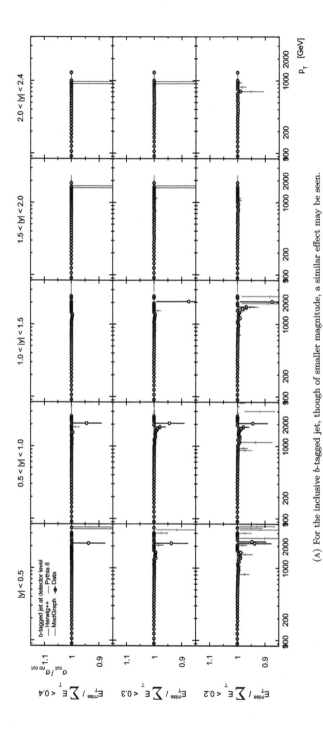

(A) For the inclusive *b*-tagged jet, though of smaller magnitude, a similar effect may be seen.

Fig. 7.20 The ratios of different samples after and before applying the different values of cut-off on the MET fraction are shown for MC and data. For the inclusive *b*-tagged jet selection, there is no effect up to 1 TeV

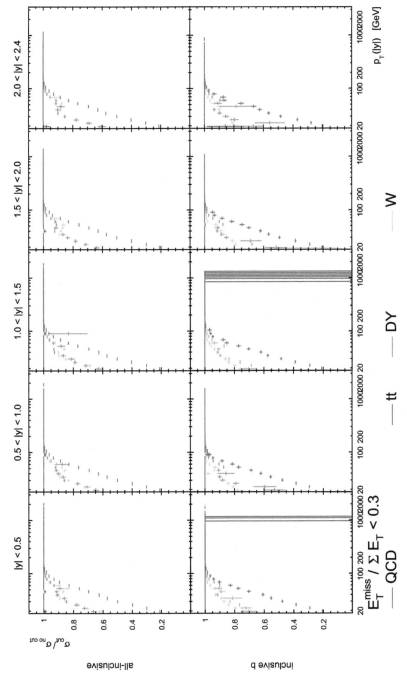

Fig. 7.21 Effect of the cut-off on the fraction of MET for the four main signals, shown for the four main contributions in the SM

Table 7.3 Turn-on points for trigger efficiency in bins of rapidity per trigger and era, shown for the two methods

Trigger	Era	Emulation method					Tag and Probe method				
		$|y|<0.5$	$0.5<|y|<1.0$	$1.0<|y|<1.5$	$1.5<|y|<2.0$	$2.0<|y|<2.4$	$|y|<0.5$	$0.5<|y|<1.0$	$1.0<|y|<1.5$	$1.5<|y|<2.0$	$2.0<|y|<2.4$
40	BCD						66	66	65	59	55
	EFearly						65	65	64	58	54
	FlateG						66	67	64	60	54
	H						65	66	65	60	56
60	BCD	77	77	77	70	66	77	77	74	69	66
	EFearly	77	78	76	69	66	77	77	74	70	61
	FlateG	77	78	77	70	68	76	77	74	69	64
	H	77	77	76	71	66	77	77	77	72	180
80	BCD	113	114	92	91	90	114	114	110	103	101
	EFearly	113	113	91	91	91	113	110	107	102	97
	FlateG	112	113	80	91	88	113	112	110	101	98
	H	110	112	92	91	91	113	109	108	102	100
140	BCD	179	179	173	164	162	179	179	173	151	162
	EFearly	177	177	171	164	162	177	177	173	150	164
	FlateG	177	179	173	165	160	177	177	171	155	164
	H	177	179	169	164	150	179	177	171	169	150
200	BCD	244	247	239	236	228	242	242	239	228	226
	EFearly	242	244	261	223	228	244	244	239	226	231
	FlateG	242	244	242	231	226	244	242	236	228	226
	H	242	242	244	223	236	236	242	236	228	226
260	BCD	311	314	304	281	294	314	311	300	281	294
	EFearly	311	307	291	284	291	314	307	294	281	291
	FlateG	314	314	291	278	284	314	311	297	287	291
	H	311	314	291	284	291	314	311	294	278	291
320	BCD	366	366	346	346	358	382	350	362	346	362
	EFearly	370	350	350	346	350	378	378	350	346	362
	FlateG	374	350	350	346	346	382	382	346	346	346
	H	350	370	350	346	350	378	374	350	346	350
400	BCD	447	447	400	447	447	452	452	447	447	447
	EFearly	447	447	442	442	447	452	452	447	447	447
	FlateG	452	447	447	442	447	452	452	447	447	447
	H	452	447	447	447	447	452	452	447	447	447
450	BCD	492	486	486	486	492	492	492	486	486	492
	EFearly	492	486	486	486	492	492	492	486	486	492
	FlateG	492	492	486	486	486	492	492	486	486	486
	H	492	492	486	486	492	492	492	486	486	492

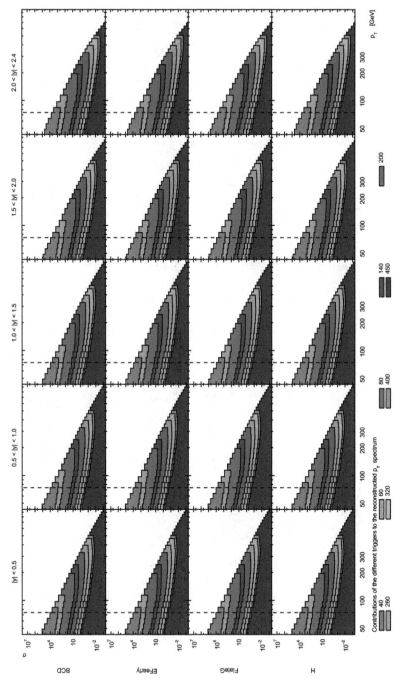

Fig. 7.22 The different contributions from the different triggers to the total cross section are stacked in bins of rapidity and per era

Table 7.4 Number of jets per trigger and per era in rapidity bins, after the final phase space subdivision

trigger	era	$|y| < 0.5$	$0.5 < |y| < 1.0$	$1.0 < |y| < 1.5$	$1.5 < |y| < 2.0$	$2.0 < |y| < 2.4$
40	BCD	868500	828016	763223	692237	488531
	EFearly	442347	419102	387150	345752	253388
	FlateG	443188	424510	391895	359266	256978
	H	445452	423063	390693	350886	253246
60	BCD	255272	243855	214973	189559	132760
	EFearly	106111	100988	89955	77792	53484
	FlateG	103910	100259	88543	78389	53386
	H	88482	85385	75478	66468	45574
80	BCD	487734	460517	401419	343303	224716
	EFearly	127217	120671	104117	88841	56976
	FlateG	123821	117553	102022	87289	55822
	H	107724	101845	88748	75023	48389
140	BCD	296982	276715	236009	192082	116147
	EFearly	102830	95836	82035	66477	39104
	FlateG	93642	86624	74113	60087	35394
	H	78884	74281	63098	50886	29596
200	BCD	633278	587602	488492	385465	221183
	EFearly	206868	191584	158455	125192	70133
	FlateG	137591	127759	106361	84264	46584
	H	116448	107935	90157	70600	39153
260	BCD	561595	514174	419367	317151	168201
	EFearly	218751	200452	163406	122542	63516
	FlateG	230859	211254	171348	129581	67287
	H	195725	179780	146313	110239	56299
320	BCD	691473	625971	501672	365205	184002
	EFearly	265658	241153	192068	139358	68677
	FlateG	285935	260335	207869	150886	73817
	H	244515	222672	177559	129016	62229
400	BCD	497249	447734	352633	246763	118903
	EFearly	200072	180630	141768	99387	46147
	FlateG	211914	191209	150252	105383	49041
	H	180553	163071	127227	90076	41326
450	BCD	5277234	4677644	3528306	2333147	1057756
	EFearly	2731499	2428308	1829205	1200886	536510
	FlateG	3300980	2929738	2214669	1465589	648822
	H	2964714	2637357	1995284	1314709	577339

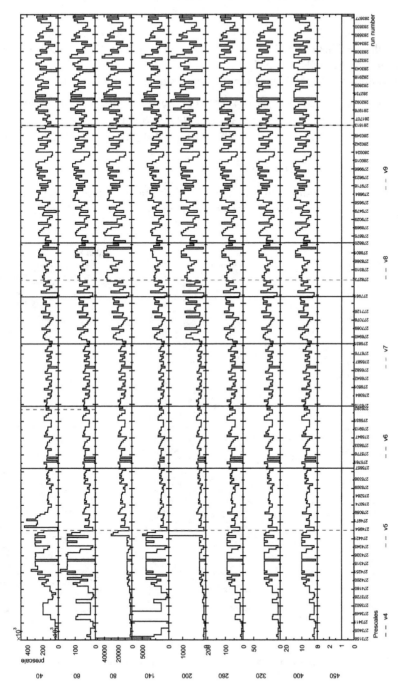

Fig. 7.23 Run-by-run averages of the pre-scales for each of the triggers used in this analysis (the threshold is indicated on the very left). 372 runs are shown, but only a few of them can be explicitly written for readability. Black vertical lines separate the different eras and the changes of version are shown by dashed lines

7.C   Additional Control Plots About Jets

Angular response. The angular response of the detector for jet reconstruction is compared in data and in the simulation in Fig. 7.24; it can be seen that the imperfections of the detector are well described.

More on binning. The following binning in considered in this analysis:

p_T 43,49,56,64,74, 84, 96, 114, 133, 153, 174, 196, 220, 245, 272, 300, 330, 362, 395, 430, 468,507, 548, 592, 638, 686, 737, 790, 846, 905, 967,1032, 1101, 1172, 1248, 1327, 1410, 1497, 1588, 1684, 1784, 1890, 2000,2116, 2238, 2366, 2500, 2640, 2787, 2941

$|y|$ 0, 0.5, 1.0, 1.5, 2.0, 2.4

Additional checks are shown in Fig. 7.25. Most importantly, the resolution for b jets is slightly worse, especially in the forward region ($|y| > 1.5$); we will see later that

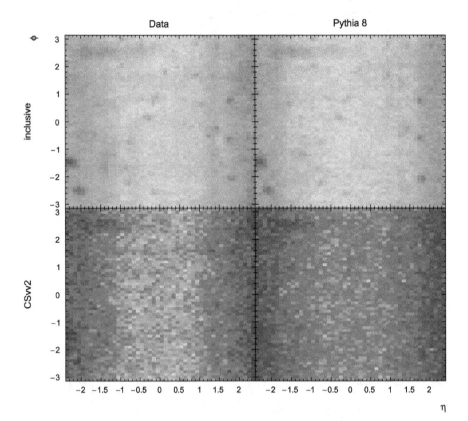

Fig. 7.24 Jet $\eta - \phi$ spectrum in data (left) and simulation (right), for inclusive jet (top) and inclusive b jet (bottom). The bins corresponds to the count of jets (not to a cross section) with arbitrary normalisation

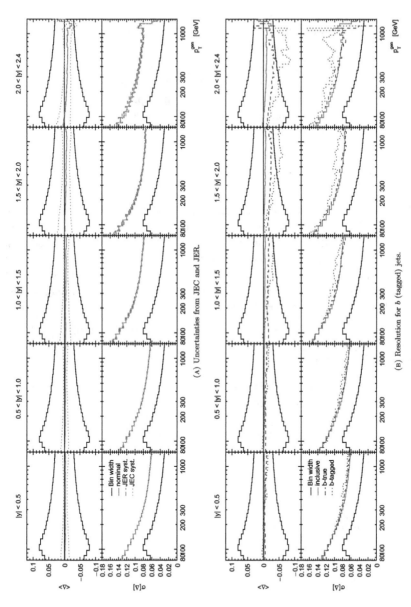

Fig. 7.25 Resolution profiles for systematic shifts and for flavoured samples

during the procedure of unfolding, bins have to be merged roughly in pairs, which will cover the migrations.

7.D More on Tagging

We show in Fig. 7.26 the performance of the tagger before calibration. In this case, one can also compare to HERWIG++: the performance is extremely similar, and the discrepancy at high transverse momentum is even more pronounced.

Additional checks may be performed by comparing taggers and WPs.

7.D.1 Comparison of the Working Points

The SFs have been shown on Fig. 7.15 for the tight selection. The medium and loose WPs are also investigated, and they can be seen on Figs. 7.27 and 7.28; the effect of different WPs on the $p_T(y)$ can be seen on Fig. 7.29. Allowing larger statistics, the SFs for the mistag of light jets can be provided with rapidity dependence in addition to transverse momentum dependence in order to attempt to mitigate the disagreement, especially in the $1.0 < |y| < 1.5$ region where it is the strongest. The rapidity dependence is defined in different binning schemes according to the WP:

– for the medium selection, it is done in three bins of width 0.8, i.e. 0.0, 0.8, 1.6, 2.4;
– for the loose selection, it is done in seven bins of different widths, i.e. 0.0, 0.3, 0.6, 0.9, 1.2, 1.5, 1.8, 2.4.

From Fig. 7.31, the fraction ratio is better described; on the other hand, the contamination from light (charm) jets reaches 60% (20%), and the purity is around 10−20% in the whole spectrum. The improvement seen in the fraction can be explained by the presence of more n jets. Therefore, we cannot conclude any improvement from the rapidity dependence of the SFs with looser WPs. The same conclusions may be drawn with the different WPs of the different taggers (not shown here).

7.D.2 Comparison of the Taggers

The different taggers can be compared in Fig. 7.32; the effect of the calibration for the different taggers on the $p_T(y)$ can be seen D on Fig. 7.30. The uncertainties from the CSVv2 (JP) tagger are the smallest (greatest) one. However, they all show similar tendencies, and confirm that possible biases are not due to their respective performances.

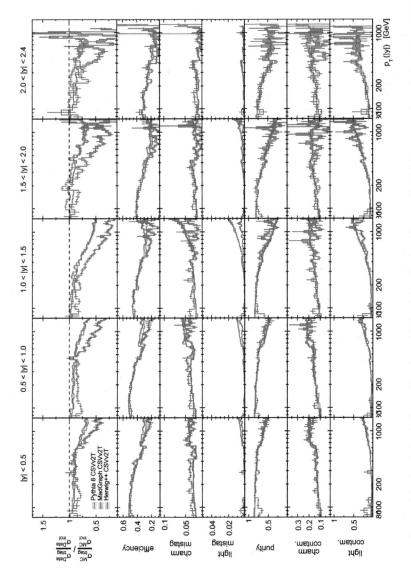

Fig. 7.26 Performance of CSVv2 in simulation before calibration with the tight selection. PYTHIA 8 and MADGRAPH share the same tune CUETP8M1 while HERWIG++ has its own tune CUETHppS1; calibration is only available for PYTHIA 8 and MADGRAPH. A significant disagreement from one in the fraction ratio can be seen for all simulations

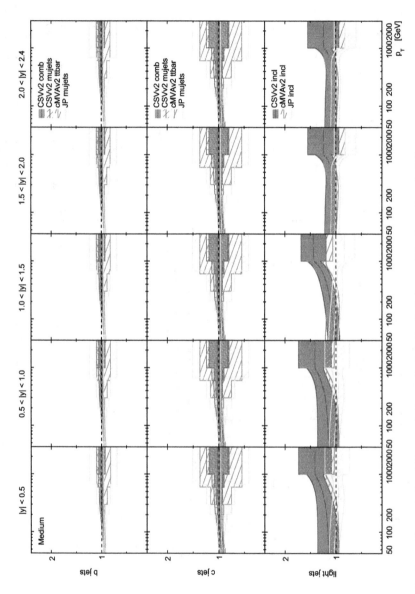

Fig. 7.27 SFs for the medium selection. Mistag scale factors for light jets include a rapidity dependence in 3 bins

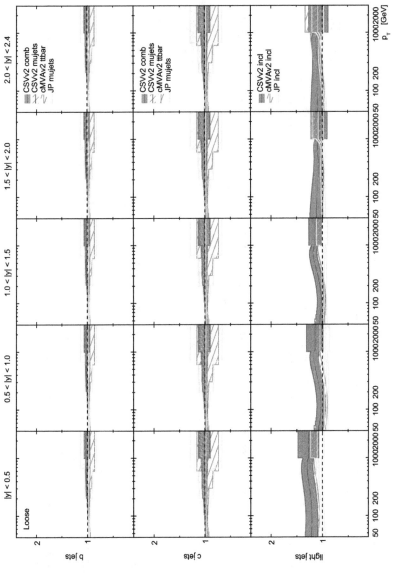

Fig. 7.28 SFs for the loose selection for the different flavours. Mistag scale factors for light jets include a rapidity dependence in 3 bins

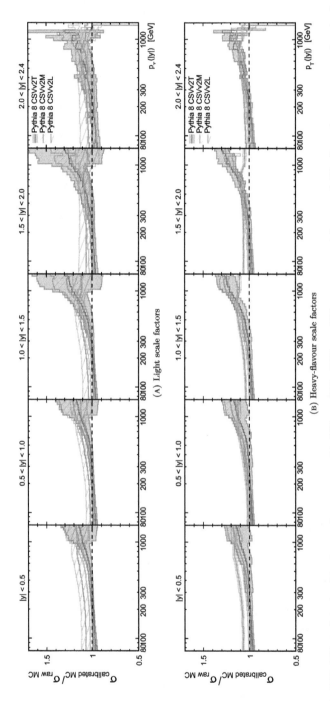

Fig. 7.29 Effect of the *b* calibration on the spectrum with different WPs. Different WPs are compared for CSVv2. The five columns correspond to the five rapidity bins. The coloured, shaded bands correspond to the uncertainties of the SFs

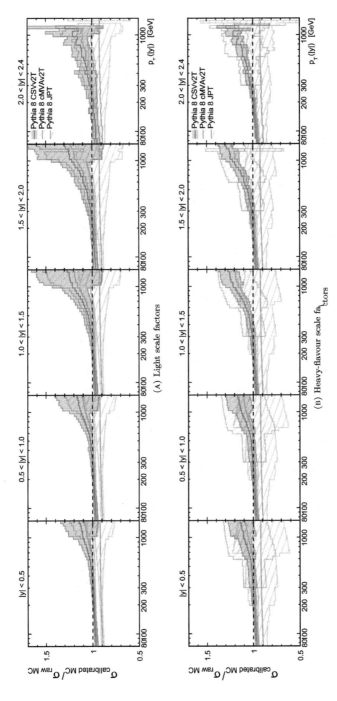

Fig. 7.30 Effect of the *b* calibration on the spectrum with different taggers. Different taggers are compared for the tight WP. The five columns correspond to the five rapidity bins. The coloured, shaded bands correspond to the uncertainties of the SFs

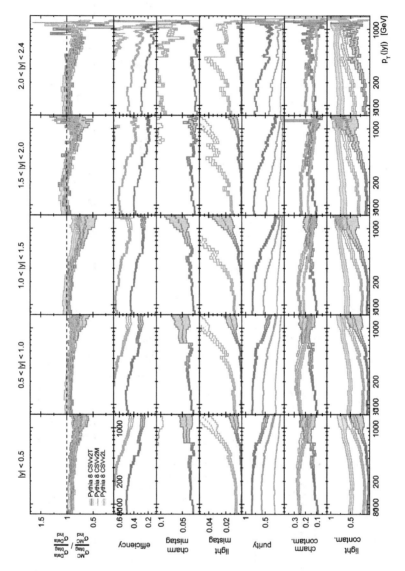

Fig. 7.31 Comparison of the performance of CSVv2 of PYTHIA 8 with different WPs

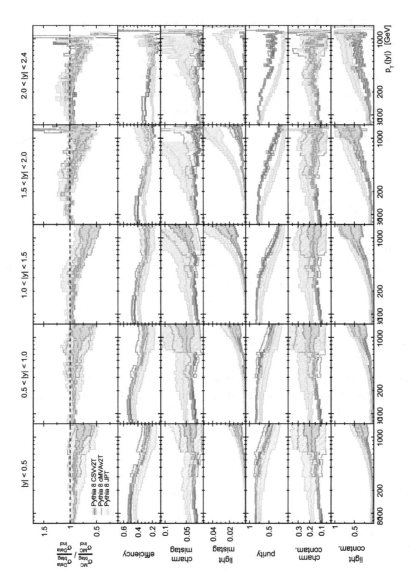

Fig. 7.32 Comparison of the performance of CSVv2 of PYTHIA 8 with different taggers

7.E Jet Constituents

The content of jets and of $\hat{b}$ jets is described by two categories of variables:

The *energy fractions* (Fig. 7.33):   The *multiplicities* (Fig. 7.34):

– charged-hadron energy fraction   – charged-hadron multiplicity
– neutral-hadron energy fraction   – neutral-hadron multiplicity
– charged e.m. energy fraction   – electron multiplicity
– neutral e.m. energy fraction   – photon multiplicity
– muon energy fraction   – muon multiplicity

(The figures for the energy fractions and for the multiplicities are shown opposite to one another.)

7.E.1 Jet ID

The *jet ID*, already addressed while describing the selection in Sect. 6.3.1, is based on these variables. In this analysis, the *tight ID* is used, to which the corresponding cut-off values for $|y| < 2.4$ are shown in Table 7.5.

7.E.2 Jet Constituents in Bins of Rapidity

In order to investigate the discrepancy in the fraction ratio (Sect. 7.4), we show some elements of additional investigations on the jet constituents in bins of rapidity.

Figures 7.33 and 7.34 show the agreement with data in bins of rapidity, after the tight jet ID selection, for inclusive jet and inclusive $\hat{b}$ jet production. Statistical uncertainties are included, but systematical uncertainties have not been investigated.

Despite the different showering used, it is interesting that PYTHIA 8 and MAD-GRAPH on the one hand and HERWIG++ and the other hand do not show any large difference on any of these variables. In general, only the variables involving neutral

Table 7.5 Tight jet ID definition in $|y| < 2.4$

PF Jet ID	Tight
Neutral hadron fraction	<0.90
Neutral e.m. fraction	<0.90
Number of constituents	>1
Charged hadron fraction	>0
Charged multiplicity	>0
Charged e.m. fraction	<0.99

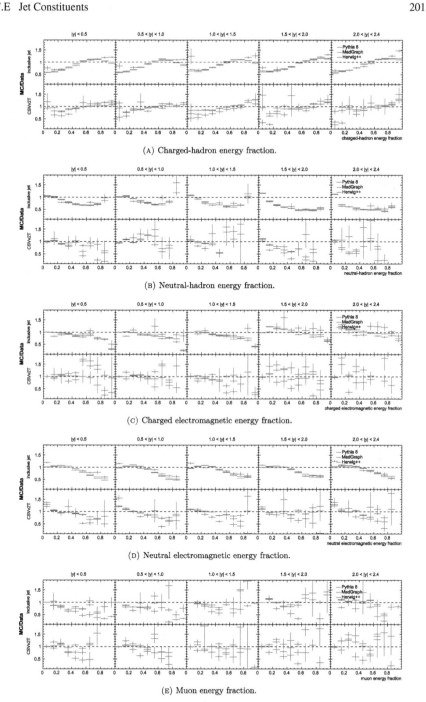

(A) Charged-hadron energy fraction.

(B) Neutral-hadron energy fraction.

(C) Charged electromagnetic energy fraction.

(D) Neutral electromagnetic energy fraction.

(E) Muon energy fraction.

Fig. 7.33 Comparison of data and simulation of the fractions of the jet constituents

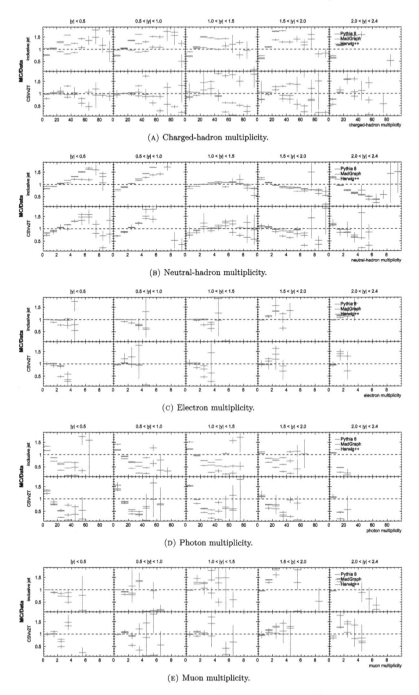

(A) Charged-hadron multiplicity.

(B) Neutral-hadron multiplicity.

(C) Electron multiplicity.

(D) Photon multiplicity.

(E) Muon multiplicity.

Fig. 7.34 Comparison of data and simulation of the multiplicities of the jet constituents

particles show a sensitive difference (Figs. 7.33b,d and 7.34b,d). But most importantly, the agreement is not affected by the b tagging.

This having been said, the statistics is usually low, and it is hard to conclude. To be perfectly rigorous in our investigations, the same investigations should be performed in bins of p_T; unfortunately, this is not possible, since the statistics are too low.

References

1. Butz E (2018) Mitigation of the strip tracker dynamic inefficiency (previously known as HIP). https://indico.cern.ch/event/560226/contributions/2277448/attachments/1324704/1988050/wgm_vfp_change_ebutz.pdf. Accessed 25 Jan 2018
2. CMS Collaboration (2013) Pileup jet identification
3. CMS Collaboration (2014) Pileup removal algorithms
4. Krohn D et al (2014) Jet cleansing: pileup removal at high luminosity. Phys Rev D 90(6):065020. https://doi.org/10.1103/PhysRevD.90.065020, arXiv:1309.4777 [hep-ph]
5. The CMS Collaboration (2011) Determination of jet energy calibration and transverse momentum resolution in CMS. J Instrum 6(11):11002 (2011). http://stacks.iop.org/1748-0221/6/i=11/a=P11002
6. CMS Collaboration (2017) CMS luminosity measurements for the 2016 data taking period
7. Salfeld J (2018) Utilities for accessing pileup information for data. https://twiki.cern.ch/twiki/bin/view/CMS/PileupJSONFileforData. CMS Private area, Accessed 25 Jan 2018
8. Cacciari M, Salam GP (2008) Pileup subtraction using jet areas. Phys Lett B 659:119–126. https://doi.org/10.1016/j.physletb.2007.09.077, arXiv:0707.1378 [hep-ph]
9. Cacciari M, Salam GP, Soyez G (2008) The catchment area of jets. JHEP 04:005 (2008). https://doi.org/10.1088/1126-6708/2008/04/005, arXiv:0802.1188 [hep-ph]
10. CMS Collaboration (2017) Jet algorithms performance in 13 TeV data
11. Khachatryan V et al (2017) Jet energy scale and resolution in the CMS experiment in pp collisions at 8 TeV. JINST 12(02):02014. https://doi.org/10.1088/1748-0221/12/02/P02014, arXiv:1607.03663 [hep-ex]
12. Sordini V (2018) Jet energy resolution. https://twiki.cern.ch/twiki/bin/viewauth/CMS/JetMET. CMS Private area. Accessed 26 Jan 2018
13. Chatrchyan S et al (2013) Identification of b-quark jets with the CMS experiment. JINST 8:04013. https://doi.org/10.1088/1748-0221/8/04/P04013, arXiv:1211.4462 [hep-ex]
14. Scodellaro L (2017) b tagging in ATLAS and CMS. In: 5th large hadron collider physics conference (LHCP 2017) Shanghai, China, May 15–20, 2017. arXiv:1709.01290 [hep-ex], https://inspirehep.net/record/1621595/files/arXiv:1709.01290.pdf
15. Sirunyan AM et al (2017) Identification of heavy-flavour jets with the CMS detector in pp collisions at 13 TeV. arXiv:1712.07158 [physics.ins-det]
16. Buskulic D et al (1993) A precise measurement of $\Gamma Z \to bb / \Gamma Z \to$ hadrons. Phys Lett B 313(3):535–548. ISSN: 0370-2693. https://doi.org/10.1016/0370-2693(93)90028-G, http://www.sciencedirect.com/science/article/pii/037026939390028G
17. Borisov G, Mariotti C (1996) Fine tuning of track impact parameter resolution of the DELPHI detector. In: Nucl Instrum Methods Phys Res Sect A: Accel, Spectrometers, Detect Assoc Equip 372(1):181–187. ISSN: 0168-9002. https://doi.org/10.1016/0168-9002(95)01287-7, http://www.sciencedirect.com/science/article/pii/0168900295012877
18. Rizzi A, Ferencek D (2018) B tagging and vertexing. https://twiki.cern.ch/twiki/bin/view/CMS/BTagSFMethods. CMS Private area, Accessed 27 Jan 2018
19. The CMS Collaboration (2011) Missing transverse energy performance of the CMS detector. In: J Instrum 6(09):09001. http://stacks.iop.org/1748-0221/6/i=09/a=P09001

20. The CMS Collaboration (2015) Performance of the CMS missing transverse momentum reconstruction in pp data at $\sqrt{s} = 8\ TeV$. J Instrum 10(02):02006. http://stacks.iop.org/1748-0221/10/i=02/a=P02006
21. Khachatryan V et al (2017) Measurement and QCD analysis of double-differential inclusive jet cross sections in pp collisions at $\sqrt{s} = 8$ TeV and cross section ratios to 2.76 and 7 TeV. JHEP 03:156. https://doi.org/10.1007/JHEP03(2017)156, arXiv:1609.05331 [hep-ex]
22. Khachatryan V et al (2016) Measurement of the double-differential inclusive jet cross section in proton - proton collisions at $\sqrt{s} = 13$ TeV. Eur Phys J C 76(8):451. https://doi.org/10.1140/epjc/s10052-016-4286-3, arXiv:1605.04436 [hep-ex]
23. Chatrchyan S et al (2012) Inclusive b-jet production in pp collisions at $\sqrt{s} = 7$ TeV. JHEP 04:084. https://doi.org/10.1007/JHEP04(2012)084, arXiv:1202.4617 [hep-ex]

Chapter 8
Analysis at Particle Level

The method of Least Squares is seen to be our best course when we have thrown overboard a certain portion of our data—a sort of sacrifice which has often to be made by those who sail the stormy seas of Probability.
— Francis Ysidro EDGEWORTH

IN THE PRESENT CHAPTER, THE EXTRACTION of the particle-level cross section from the measurement is described.

The chapter is organised as follows. First, the disagreement in the b-tagged fraction observed in the previous chapter is investigated; a correction to the simulation is applied to fix the disagreement. Then, the b-jet and n-jet cross sections are simultaneously extracted together with advanced techniques of unfolding; the treatment of the systematic uncertainties in the unfolding is also discussed.

8.1 Purity

The CMS taggers were described in Sect. 7.4; the performance was given. The efficiency and mistag rates, as well as the purity and the contamination, are two equivalent ways to describe the effect of tagging from different point of views: the former (latter) describes the content of the tagged (true) sample in terms of true (tagged) sample:

– from true to tagged:

$$\begin{bmatrix} \sigma_{\hat{n}} \\ \sigma_{\hat{b}} \end{bmatrix} = \begin{bmatrix} 1 - m & 1 - \epsilon \\ m & \epsilon \end{bmatrix} \times \begin{bmatrix} \sigma_n \\ \sigma_b \end{bmatrix} \tag{8.1}$$

where ϵ (m) is the *efficiency* (*mistag*);

© Springer Nature Switzerland AG 2019
P. Connor, *Inclusive b Jet Production in Proton-Proton Collisions*, Springer Theses,
https://doi.org/10.1007/978-3-030-34383-5_8

– or *vice-versa*:

$$\begin{bmatrix} \sigma_n \\ \sigma_b \end{bmatrix} = \begin{bmatrix} P_{n\hat{n}} & 1 - P_{b\hat{b}} \\ 1 - P_{n\hat{n}} & P_{b\hat{b}} \end{bmatrix} \times \begin{bmatrix} \sigma_{\hat{n}} \\ \sigma_{\hat{b}} \end{bmatrix} \tag{8.2}$$

where $P_{b\hat{b}}$ ($P_{b\hat{b}}$) is the *purity* of the tagged (non-tagged) sample.

The purity and contaminations are related to the efficiency and mistag rates by matrix inversion. Therefore, the b calibration described in the previous chapter corrects the purity of the samples at detector level.

The extraction of the true cross section can in principle be performed by matrix inversion; technically, this operation from tagged to true cross section is included in the unfolding.

In Fig. 7.17, a strong disagreement the fraction of b-tagged jet in the inclusive jet sample was observed between simulation and data, despite the calibration. In this section, we discuss origin of the disagreement and show how a correction on the purity can fix it.

8.1.1 Origin of the Disagreement

At high p_T, in addition to the disagreement in the fraction ratio, one can observe that the efficiency and the purity (mistag rate and contamination) are significantly low (high).

In principle, after the calibration of the b-tagging, the efficiency and mistag rates are supposed to be correct; however, at high p_T, the SFs are not derived in the inclusive-jet sample, where the fraction of b jets is small, but from sub-samples where the statistics of b and n jets are of similar order. Consequently, in the present analysis, a tiny under- or overestimation of the performance can lead to important differences between data and simulation in the tagged sample. For instance, numerically, Eq. 8.1 reads:

$$\begin{bmatrix} \sigma_{\hat{n}} \\ \sigma_{\hat{b}} \end{bmatrix} = \begin{bmatrix} 0.99 & 0.5 \\ 0.01 & 0.5 \end{bmatrix} \times \begin{bmatrix} \sigma_n \\ \sigma_b \end{bmatrix} \tag{8.3}$$

Since $\sigma_n \approx 50\sigma_b$, a variation of the order of the percent is enough to induce significant effects in the tagged cross section; if this is not correct at the level of the detector, it infers at particle level.

Moreover, the treatment of rapidity dependence in the training and in the calibration of CSVv2 is not satisfactory: only two bins to train the tagger are considered, and no rapidity dependence is included in the calibration. The high p_T region is also treated with a very rough binning (see Fig. 7.2), which is not so fine as the one used in the current analysis.

Therefore, an additional correction has to be derived.

8.1.2 Method of the Determination of the Purity

In Fig. 8.1, the two first sub-figures correspond to the procedure of calibration described in the previous chapter, while the third corresponds to the additional correction that is the object of this section.

Fig. 8.1 Sketch of the calibration. The three sub-figures represent the CSVv2 tagger at different stages of the calibration. The blue (red) stands for the light (*b*) jets; *c* jets are not represented

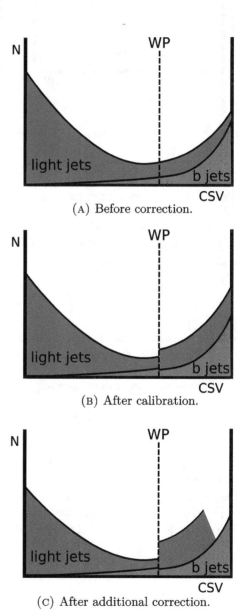

(A) Before correction.

(B) After calibration.

(C) After additional correction.

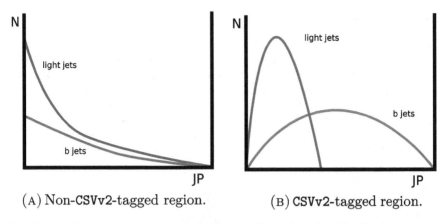

(A) Non-CSVv2-tagged region.   (B) CSVv2-tagged region.

Fig. 8.2 Sketch of the JP templates in the two regions. Charm is neglected in this picture

The principle consists in performing a *template fit* of the purity separately in the $\hat{b}$ and $\hat{n}$ samples, using a variable sensitive to b jets but loosely correlated with the CSVv2 tagger. A good choice for such a variable is for instance the JP tagger (see correlations in Sect. 7.4 in Fig. 7.13).

In a template fit, rather than an analytical function, one uses *templates* t_is (for instance obtained from the simulation) to define a function:

$$f(x) = \sum_i p_i t_i(x) \tag{8.4}$$

where the p_is are the fit parameters. The technique of template fits is standard when the analytical shape of a function is unknown. Here, a template fit of the discriminant is performed in the data $N_{\text{data}}^{\text{total}}$ with templates corresponding to the discriminant for the different flavours in simulation $N_{\text{MC}}^{\text{flavour}}$:

$$N_{\text{data}}^{\text{total}}(\text{JP}) = \sum_{\text{flavours}} p_{\text{flavour}} N_{\text{MC}}^{\text{flavour}}(\text{JP}) \tag{8.5}$$

where p_{flavour} are the fit parameters. Figure 8.2 illustrates the templates of JP for the different flavours in the CSVv2-tagged and non-CSVv2-tagged sub-samples. After the fit, the contributions to the CSVv2 distribution are renormalised.

A priori, one could simply perform a template fit of the CSVv2 tagger itself; however, using another variable results in a better fit since the power of discrimination of JP is used in addition to the one of CSVv2. Also, in the CSVv2-tagged region, the JP templates have very distinct shapes (as illustrated in Fig. 8.2) whereas the different flavours have similar shapes for CSVv2; the fit has more chances to converge with distinct shapes.

8.1.3 Results

The fits are performed in bins of transverse momentum and rapidity. The step-by-step procedure is described in details in Sect. 8.3.

In practice, in order to make the fit converge, we have to face two difficulties:

1. In the whole discussion till now, the special behaviour of c jets has been neglected. However, in practice, the template for the c component has a shape halfway between light and b templates. Different alternatives can be thought of: either the three templates are considered independently, or the c template is constrained together with one of the other templates (either the light or the b template). The most stable configuration happens when the b and c templates are constrained together. In order to cover this arbitrary constrain, a conservative uncertainty on the procedure is derived by varying the normalisation of the c template with a factor of 1.0 ± 0.5:

$$N_{\text{data}}^{\text{total}} = p_{\text{udsg}} N_{\text{MC}}^{\text{udsg}} + p_{\text{b+c}} \left((1.0 \pm 0.5) N_{\text{MC}}^{c} + N_{\text{MC}}^{b} \right) \qquad (8.6)$$

2. Moreover, in the non-CSVv2-tagged region, the statistics from the light and c components is much larger than the b component's; for this reason, the template fit in this region did not converge or give satisfactory results. As an alternative, we apply the same normalisation factor to b and c jets as obtained in the CSVv2-tagged region, and simply rescale the light component [1].

The simulation is corrected to match the purity in data. Figure 8.3 shows the performance of the CSVv2 tagger in PYTHIA 8 before and after the applying the correction. The performance of the CSVv2 tagger is improved by the procedure: one can see that the fraction ratio agrees at one and that the mistag rate is higher at high p_T after the correction than before.

8.2 Unfolding

The aim of unfolding is to correct for different effects of the detector on the measurement of a differential distribution [2]. These effects are the limited resolution, acceptance, inhomogeneities and anisotropies of the detector. They can result in a smearing due to migrations or in reconstruction inefficiencies.

The problem of unfolding may be formulated in the following mathematical terms:

$$\mathbf{A}\mathbf{x} + \mathbf{b} = \mathbf{y} \qquad (8.7)$$

where

- $\mathbf{x}$ is the *truth* vector at particle-level, which represent the distribution that we want to determine in this section;

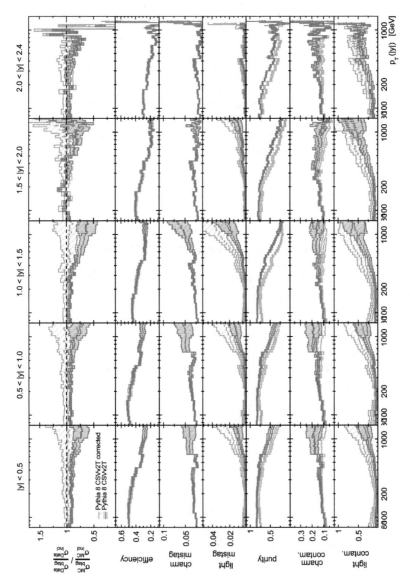

Fig. 8.3 Comparison of the performance of b-tagging before (red) and after (green) the additional correction. The columns correspond to the rapidity bins; the rows correspond to the ratio of the fractions in data and simulation (first row), to the efficiency and mistag rate (three next rows) and to the purity and contaminations (three last rows)

- **y** is the *measurement* vector at detector-level, which is known from Chap. 7;
- **b** is the *background* vector at detector-level, discussed in Chap. 6;
- and **A** is the Probability Matrix (PM), which we are going to construct with the MC samples, where an element A_{ij} corresponds to the probability that a quantity with true value (or generated values for MC) in bin j will be measured (or reconstructed) in bin i.

Given **y**, **b** and **A**, the exact solution $\mathbf{x} = \mathbf{A}^{-1}(\mathbf{y} - \mathbf{b})$ is not guaranteed to work due to numerical instabilities, both in the measurement and in the PM. Therefore the matrix inversion is replaced with the so-called *unfolding*; here, we mainly use the unfolding formulated as a least square minimisation with Tikhonov regularisation, implemented with the TUnfold package (version 17.6) [3, 4].

In this section, the construction of the response of the detector and the technique of unfolding are first described. The unfolding is repeated in different conditions:

- we perform a Closure Test (CT) on pseudo-data with the simulations;
- we perform the unfolding with two different MC samples and with two different taggers;
- we cross-check the result with another unfolding technique.

The treatment of the uncertainties is explained along the different steps. Additional tests like the *backfolding* and Bottom Line Test (BLT) are presented in Sect. 8.4.

8.2.1 Response of the Detector

In this subsection, we describe how the response of the detector is implemented in the Response Matrix (RM), and what the Response Matrix (RM) describes in general and in the particular case of the current analysis.

8.2.1.1 Response Matrix and Probability Matrix

A RM consists in a matrix that contains the information both at particle level and detector level. A RM can be constructed for instance from a MC sample containing event records at both levels, e.g. PYTHIA 8 or MADGRAPH.

In the case of the unfolding of, say, the p_T spectrum, the RM is a 2-dimensional matrix, with one coordinate corresponding to each level; it is obtained from the simulation by filling a 2-dimensional histogram with pairs $(p_T^{\text{rec}}, p_T^{\text{gen}})$ corresponding to the values of the transverse momentum of a jet before and after the simulation of the detector. The correspondence between particle level and detector level is however not obvious, since a particle-level jet may be reconstructed as two separated detector-level jets or *vice-versa*. One needs to define a *matching* between the jets of the two levels to obtain the pairs $(p_T^{\text{rec}}, p_T^{\text{gen}})$; the matching is described in one of the next paragraphs (Sect. 8.2.1.2).

As already mentioned in Sect. 6.3.4 while discussing the strategy of the analysis, we consider in the present analysis three variables simultaneously in the unfolding:

1. the transverse momentum,
2. the absolute rapidity,
3. and the flavour.

The unfolding will be 3-dimensional, and RM 6-dimensional, but the principles are the same: jets are matched between the two levels, and fill the RM with a couple of values for each variable.

The RM from PYTHIA 8 is shown in Fig. 8.4: the four large sectors correspond to the flavour components, similarly to Eq. 8.1; inside of each of the sectors are the 5×5 rapidity cells; finally, each rapidity cell consists of a simple 2-dimensional RM for the transverse momentum. Inside of a cell, the entries are concentrated around the diagonal, since jets are most likely reconstructed at the same value as at particle level; moreover, the dispersion on either side of the diagonal is slightly asymmetric, since it is more likely to miss some elements of a jet at the reconstruction than including extra activity. These two facts were already observed in Sect. 7.3; in Fig. 7.11, the mean value of the resolution profile is smaller than the bin width (which here translates in being concentrated around the diagonal) but is not at zero (which reflects the asymmetry of the dispersion). That the RM is mostly diagonal is important to perform the unfolding, as will be discussed later (Sect. 8.2.2.1).

The binning schemes of the generated and reconstructed axes differ only for the p_T component, where bins are merged by two (or more) with respect to the binning described in Sect. 7.3; merging bins is part of the regularisation, as will be defined later in this section (Sect. 8.2.2.2).

So far, we have described the construction of the RM; but in Eq. 8.7, the unfolding is performed with the PM. The PM $\mathbf{A}$ is obtained by normalising every column of the RM $\hat{\mathbf{A}}$ to the total number of the generated events in the corresponding bin $\mathbf{x}^{MC}$ at particle level:

$$A_{ij} = \frac{1}{x_j^{MC}} \sum_i \hat{A}_{ij} \tag{8.8}$$

As some jets are not reconstructed, the sum of elements in every column of the PM is smaller than 1; this value corresponds to the efficiency of reconstruction for the given bin at particle level. We discuss in the next sections the migrations among bins and the inefficiencies of reconstruction.

8.2.1.2 Matching

To build up the RM from the MC samples, we adopt the following procedure to match jets from detector level to particle level:

1. Reconstructed jets are considered one by one from the highest to the lowest transverse momentum.

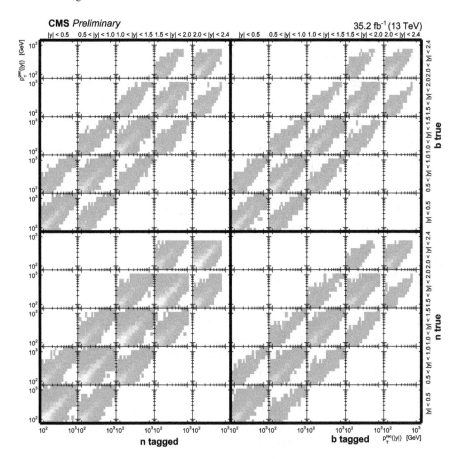

Fig. 8.4 Six-dimensional RM extracted from the PYTHIA 8 sample: migrations are shown for p_T (smallest cells), $|y|$ (intermediate-size cells) and the flavour (large sectors). The vertical (horizontal) axis corresponds to the generated (reconstructed). The cells contain event counts, up to $\sim 10^2$ (yellow) and down to $\sim 10^{-13}$ (blue)

2. Generated jets are considered around the reconstructed jet in a cone of radius $\Delta R = \sqrt{\phi^2 + y^2} = 0.2$. It was checked that the description of the migrations is not sensitive to variations of the radius $\Delta R \in [0.15, 0.40]$ (not shown here).

3. Inside of the cone, the generated jet of highest transverse momentum is considered for the matching, unless it was already matched previously. Another strategy is also possible, by considering the closest generated jet instead of the one of highest transverse momentum, but this has a negligible impact.

The matching is not always defined with a cone, and is sometimes only based on p_T ordering. However, this would be sensitive to additional jet activity.

By definition of the matching condition, there cannot be migrations among non-neighbouring rapidity bins; this explains the empty bins of middle size in the top left and bottom right corners of the flavour sectors of the RM in Fig. 8.4.

8.2.1.3   Miss and Fake Jets

The generated (reconstructed) , non-matched jets are called *miss* (*fake*) jets:

– The miss jets are jets that were not matched, most often because they were reconstructed outside of the measured phase space. The miss jets are treated as yet
  another possible bin in the RM (although not represented in Fig. 8.4).
– The fake jets are jets that were reconstructed in addition; several reasons can hold
  for this: it can come from additional activity (pile-up or MPI); it can be due to
  a bad clustering of the jet; it can also be related to difficulties of reconstructing
  jets in certain regions of the detector (e.g. in presence of a gap or of a joint). The
  fake jets are considered as a background; in fact, this is the only background in
  the analysis, and it only affects the bins at the edge of the phase space, i.e. the two
  or three first p_T bins and the last rapidity bin.

8.2.1.4   Description of the Migrations

Figure 8.4 helps visualising the migrations globally. In complement, the composition of the bins of p_T and y in terms of migrations is shown in Fig. 8.5 for $\hat{n}$ and
$\hat{b}$ jets. Except the green, which stands for jet generated and reconstructed in the same
(p_T, $|y|$, flavour) bin, each colour corresponds to a type of migration.

   We can describe the migrations in terms of *purity* (not in the restricted sense of
the flavour):

– the non-b-tagged-jet cross section at detector level shows mainly migrations in
  p_T, and has a purity above 60%;
– however, the b-tagged-jet cross section suffers also from migrations due to the
  tagging, and the purity goes down to around 20% at $p_T = 1$TeV.

The gray area corresponds to the fake jets, i.e. the background, and shows that the first
p_T bin has typically a background of 30%, rapidly decreasing to 10% in the second
bin and getting negligible in bins at higher values. An uncertainty is extracted by
scaling up and down the subtracted background with 10%; this is a rough but very
conservative estimation, since the matching is very robust against variations of ΔR.

8.2.1.5   Uncertainties

In Chap. 7, the following systematic uncertainties have been discussed:

– pile-up reweighting (Sect. 7.2)
– pile-up removal (Sect. 7.2)
– JER smearing (Sect. 7.3)
– JES correction (Sect. 7.3)
– b tagging calibration (Sect. 7.4)
– b purity (Sect. 8.1)

In addition, we mentioned the background subtraction in Sect. 8.2.1.3.

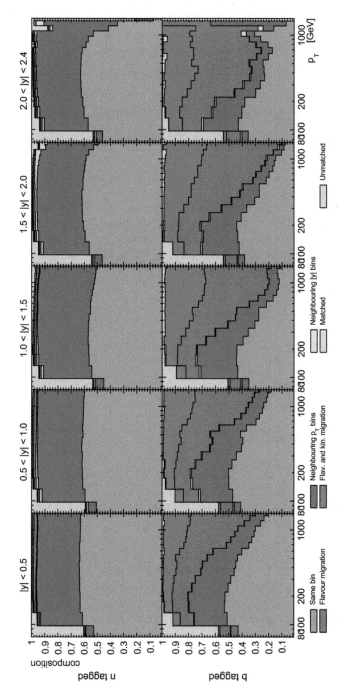

Fig. 8.5 Migrations of the spectrum from generator level to detector level. The upper (lower) row shows the $\hat{n}$ ($\hat{b}$) jets; the rows (columns) correspond to flavour (rapidity bin). Each colour (except green) corresponds to a different type of migration: blue (yellow, purple) corresponds to variation of p_T (|y|, flavour) only from neighbouring bins; red (cyan) corresponds to simultaneous p_T and |y| migrations in neighbouring (non-neighbouring) bins. Finally, fake jets are shown in gray

The RM is constructed separately for each variation up and down. The systematics uncertainties are therefore inferred at particle level by repeating the unfolding for each variation of the RM.

In addition, the RM itself has statistical uncertainties that contribute to the unfolded spectrum as an additional uncertainty.

8.2.1.6   Model Reweighting

A last procedure is performed to improve the construction of the RM. Since the models can have distinct distributions from the data, the unfolding can infer some significant bias in the data. The model reweighting aims at reducing the model dependence of the unfolding.

In the present case, the model reweighting is performed together with the correction of the flavour purity: the fit of the purity is performed without normalisation of the fitted discriminant; then the fit parameters are directly applied to the template, which results in corrected simulations describing the data. Since the flavour is fitted in bins of p_T and y, the spectrum in the simulation is corrected to the one in the data also in terms of kinematics.

8.2.2   Procedure

We have now determined the different components entering Eq. 8.7:

– the measurement at detector level **y** from Chap. 7,
– the background **b**
– and the PM **A** from Sect. 8.2.1.

Therefore we can now describe the procedure of unfolding.

In principle, the unfolding consists in the inversion of **A** (or more generally pseudo-inversion, as the RM is not necessarily square). In practice however, the inversion of the matrix is often numerically unstable and may not lead to reasonable results, and may even deliver negative values.

In this subsection, the origin of the instabilities are explained, and the strategy to prevent instabilities from degrading the final result, the *regularisation procedure*, is described.

8.2.2.1   Origin of the Instabilities

The origin of fluctuations can be understood in terms of Singular Value Decomposition (SVD) [5]:

$$\mathbf{A} = \mathbf{USV}^\mathsf{T} \tag{8.9}$$

where $\mathbf{U}$ and $\mathbf{V}$ are orthogonal matrices transforming $\mathbf{A}$ into $\mathbf{S}$, a square matrix with the eigenvalues s_i of $\mathbf{S}$ on the diagonal. The inverse of $\mathbf{S}$ depends on the inverse of the determinant of the matrix $|\mathbf{S}| = \Pi s_i$; if the eigenvalues differ in order of magnitude, the inversion of the PM is dominated by small eigenvalues. The *regularisation* consists of limiting the contributions from the small eigenvalues.

A relevant quantity quantifying whether the matrix inversion is possible (or in other words, how much regularisation is necessary) is the *condition* of the matrix, which is the ratio of the highest eigenvalue with the lowest eigenvalue: if they are of different orders of magnitude, the *condition* of the matrix should be very different from 1. In this analysis, the *condition* of the PM from PYTHIA 8 is 13.586, therefore no large impact of the regularisation is expected.

8.2.2.2 Tikhonov Regularisation

The regularisation chosen for the present analysis is the *Tikhonov regularisation* [5].

The problem of unfolding is reformulated as a problem of least-square minimisation:

$$\chi^2 = (\mathbf{y} - \mathbf{Ax})^T \mathbf{V}_y^{-1} (\mathbf{y} - \mathbf{Ax}) \tag{8.10}$$

where $\mathbf{V}_y$ is the covariance matrix from the measurement (shown in Sect. 8.4). As the inclusive b jet measurement is a multi-count observable, off-diagonal elements are different from zero.

Minimising χ^2 naturally leads to a good estimate of the invert of the PM $\mathbf{A}$ within the available statistics.

Two types of regularisation are applied:

- The number of free parameters to fit is reduced; in other words, a finer binning is used for the measurement than for the final result at particle level. It should first be noted that the perfect solution, $\mathbf{x} = \mathbf{A}^{-1}\mathbf{y}$, using a square RM, leads to $\chi^2 = 0$. On the other hand, the opposite case of a single-row matrix does not necessarily lead to $\chi^2 = 1$, which means that, ideally, the number of degrees of freedom should be tuned to perform the unfolding. However, no attempt has been done to find an optimal merging scheme; the recommendation in the TUnfold package is to perform the measurement with twice more bins than at particle level has been followed [6].
- A second term is added in the χ^2 in Eq. 8.10, the *regularisation term*:

$$\chi^2 = \underbrace{(\mathbf{y} - \mathbf{Ax})^T \mathbf{V}_y^{-1} (\mathbf{y} - \mathbf{Ax})}_{\chi_A^2} + \tau^2 \underbrace{||\mathbf{Lx}||^2}_{\chi_L^2} \tag{8.11}$$

where $\mathbf{L}$ is the *regularisation matrix* and τ the *regularisation parameter*. In general, this additional term aims at constraining close bins to one another. The form of the regularisation matrix is specific to each unfolding problem and its choice will be detailed in the next section. The regularisation parameter balances the contribution of the two terms to the final result; the choice of the regularisation has to be performed according to objective criteria and will be detailed later in the second next section.

In the present case, only migrations related to the transverse momentum may need to be regularised:

- The rapidity spectrum is rather flat, and the resolution is finer than the rapidity binning; no regularisation needs to be applied in that case.
- The flavour determination, as discussed in the previous section, is only performed in two bins; no regularisation can even be applied.
- In contrast, the transverse-momentum spectrum is steeply falling, and the resolution is of the order of the bin sizes; therefore, significant migrations cause the smearing of the spectrum.

8.2.2.3 Regularisation Matrix

Different prescriptions exist to build the regularisation matrix (or $\mathbf{L}$ matrix) [3]: here, we choose the prescription consisting in minimising the second derivative of the cross section with respect to p_T, i.e. in relating any three adjacent bins in p_T so as to prevent fluctuations from getting significant; this is called the *curvature regularisation*. A default regularisation mode is available in TUnfold; however, in the present case, we will see that it is not convenient and that we need to modify it.

In mathematical terms, the curvature regularisation means that, given any three consecutive bins j_1, j_2, j_3, we minimise the following quantity $(x_{j_3} - x_{j_2}) - (x_{j_2} - x_{j_1}) = x_{j_3} - 2x_{j_2} + x_{j_1}$ where the x_is are the true values (the unknowns); this quantity can be seen as the numerical second derivativeThe $\mathbf{L}$ matrix is then:

$$\mathbf{L} = \begin{bmatrix} 0 \ldots 0 & 0 & 1 & -2 & 1 \\ 0 \ldots 0 & 1 & -2 & 1 & 0 \\ 0 \ldots 1 & -2 & 1 & 0 & 0 \\ \vdots & \vdots & \vdots & \vdots & \vdots \\ 0 & 0 & 0 & 0 & 0 \end{bmatrix} \tag{8.12}$$

Note that as one constraint binds three consecutive p_T bins, there are two less constraints (rows) than there are p_T bins (columns).

Because of the steeply falling character of the cross section with respect to p_T, Eq. 8.12 needs to be modified, i.e. in order for the high-p_T bins to be numerically significant in the minimisation, an additional factor may be added to build up the matrix so that all terms in χ_L^2 are of similar order of magnitude. In terms of formula, this means that we want to introduce weights ms in the second derivative $m_a^j(x_{j_3} - x_{j_2}) - m_b^j(x_{j_2} - x_{j_1})$; the $\mathbf{L}$ matrix in Eq. 8.12 becomes:

$$\mathbf{L} = \begin{bmatrix} 0 \cdots & 0 & 0 & m_a^j & -m_a^j - m_b^j & m_b^j \\ 0 \cdots & 0 & m_a^k & -m_a^k - m_b^k & m_b^k & 0 \\ 0 \cdots & m_a^l & -m_a^l - m_b^l & m_b^l & 0 & 0 \\ \vdots & \vdots & \vdots & \vdots & \vdots \\ 0 & 0 & 0 & 0 & 0 \end{bmatrix} \tag{8.13}$$

For the current analysis, we simply use the MC prediction at particle level $\mathbf{x}^{\mathrm{MC}}$, which means that $m_a^j = \frac{1}{x_{j_3}^{\mathrm{MC}}}$ and $m_b^j = \frac{1}{x_{j_1}^{\mathrm{MC}}}$. The matrix then looks as follows:

$$\mathbf{L} = \begin{bmatrix} 0 \cdots & 0 & 0 & \frac{1}{x_{j_3}^{\mathrm{MC}}} & -\frac{1}{x_{j_3}^{\mathrm{MC}}} - \frac{1}{x_{j_1}^{\mathrm{MC}}} & \frac{1}{x_{j_1}^{\mathrm{MC}}} \\ 0 \cdots & 0 & \frac{1}{x_{k_3}^{\mathrm{MC}}} & -\frac{1}{x_{k_3}^{\mathrm{MC}}} - \frac{1}{x_{k_1}^{\mathrm{MC}}} & \frac{1}{x_{k_1}^{\mathrm{MC}}} & 0 \\ 0 \cdots & \frac{1}{x_{l_3}^{\mathrm{MC}}} & -\frac{1}{x_{l_3}^{\mathrm{MC}}} - \frac{1}{x_{l_1}^{\mathrm{MC}}} & \frac{1}{x_{l_1}^{\mathrm{MC}}} & 0 & 0 \\ \vdots & \vdots & \vdots & \vdots & \vdots \\ 0 & 0 & 0 & 0 & 0 \end{bmatrix} \tag{8.14}$$

8.2.2.4 Regularisation Parameter

The `TUnfold` package offers several methods to determine the regularisation parameter τ [3]:

- the L-curve scan,
- the minimisation of the global correlation coefficients.

Here, we use the L-curve scan method: the so-called L-curve is defined by the pairs (L_x, L_y), depending on the parameter τ:

$$L_x = \log_{10} \chi_{\mathbf{A}}^2 \tag{8.15}$$
$$L_y = \log_{10} \chi_{\mathbf{L}}^2 \tag{8.16}$$

where χ_A^2 (χ_L^2) corresponds to the first (second) term of the solution in Eq. 8.11. The L-curve has a L-shape (hence its name), and the value of τ is then chosen at the point of highest curvature ("in the L"). This choice gives a good compromise in the minimisation of the two terms, where χ_A^2 does not increase (which means that the agreement of $\mathbf{Ax}$ and $\mathbf{y}$ is not degraded) while χ_L^2 is minimised (which means that the fluctuations are limited).

The L-curve is shown in Sect. 8.2.2.6, while performing the unfolding in data.

An uncertainty on the regularisation is then obtained by varying the regularisation parameter τ (but the regularisation matrix $\mathbf{L}$ is kept identical).

8.2.2.5   Closure Test

The Closure Test (CT) is performed to validate the procedure of unfolding and the MC samples. It consists in unfolding pseudo-data, i.e. a simulation where the truth is already known. Then, the unfolded and true spectra can be compared.

The CT can be seen in Fig. 8.6; in this figure, three levels of CTs are presented:

– First, we unfold PYTHIA 8 with itself (in red in the figure); in that case, the unfolded spectrum and the generated curve are identical and the agreement is therefore at one.
– Then, we split the PYTHIA 8 into two statistically orthogonal sub-samples; one is used as pseudo-data and the other as MC to construct the RM (yellow). This way, statistical correlations are avoided, but the pseudo-data is *per se* perfectly modelled. The agreement is not exactly at one but around is for all values of p_T.
– Finally, we unfold MADGRAPHwith PYTHIA 8 (blue), where, despite the model reweighting, remaining differences are still included. In this case, one observe more fluctuations of the order of a few percents, and small deviations at high p_T.

8.2.2.6   Results in Data

We consider the unfolding with the three scenarios in parallel, in order to cross-check the choice of the MC sample and of the tagger. The three scenarios correspond to the following:

– PYTHIA 8 using CSVv2 (in red in Figs. 8.7, 8.8, 8.9 and 8.10)
– PYTHIA 8 using cMVAv2 (in orange)
– MADGRAPHusing CSVv2 (in blue)

We perform the unfolding with TUnfold, including all uncertainties described in Sect. 8.2.1.5 and the variation of the regularisation parameter.

L-curve scan. The L-curve scan is performed by minimising Eq. 8.11 200 times consecutively for different values of the parameter τ in the range $[1 \times 10^{-4}, 3 \times 3 \cdot 10^{-1}]$. The L-curve scan is presented in Fig. 8.7 in either scenario: in the all cases, the regularisation is weak, and small values of τ are taken (therefore the L-shape of the curve may be difficultly to recognise). In addition, the two contributions to the total χ^2 are shown as a function of τ: one can then see that τ is taken before the χ_A^2 increases.

Cross section. The ratio of the double differential cross sections for b jets and n jets of data with the three simulations is shown in Fig. 8.8 (8.9) before (after) unfolding. The two figures are shown opposite and are similarly organised, with the flavour (rapidity) in the rows (columns). Before unfolding, since the simulations have been reweighted to describe the data (see Sect. 8.2.1.6), we see that the simulation agrees indeed well with the data, except for low and large p_T values for the $\hat{n}$ bins in MADGRAPH. After unfolding, in Fig. 8.9, one can observe the same disagreement; it should be noted

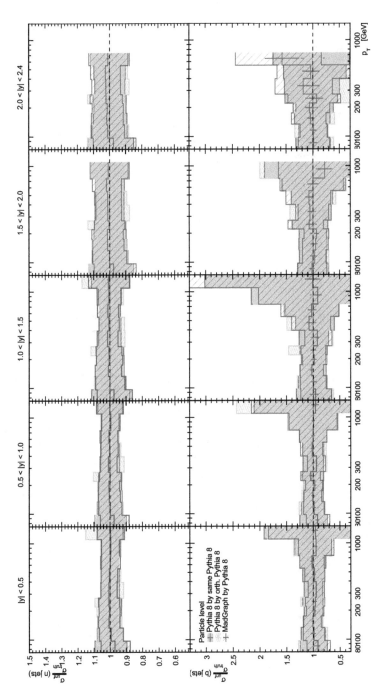

Fig. 8.6 The three CTs are shown. The rows (columns) correspond to the flavour (rapidity). The ratio is taken between the unfolded curve and the generated curve. The bands correspond to the statistical and systematic uncertainties

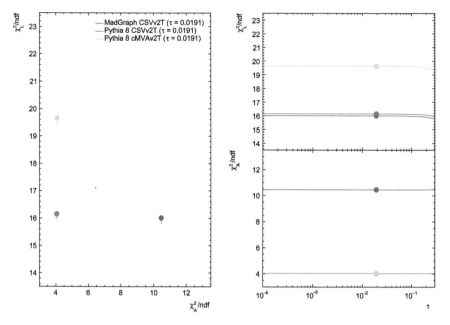

Fig. 8.7 The L-curves are represented on the left hand side for the different scenarios. The χ^2_A and χ^2_L are plotted as a function of τ on the right hand side. The points represent the choice of τ

that each simulation is considered separately: the ratio after unfolding with a given MC is taken with the same MC.

Comparison of the scenarios. The comparison of the unfolded curves in the three different scenarios is shown in Fig. 8.10, where PYTHIA 8 is taken as reference. The figure is organised in the same way as the figure for the cross sections, with the rows (columns) corresponding to the flavour (rapidity). The disagreement of MADGRAPHwith data in the $\hat{n}$ cross section at detector level does not play a rôle in the b jet cross section; however, PYTHIA 8 with cMVAv2 exhibits differences at high p_T up to 20% in the third rapidity bin, though still within the uncertainties.

Uncertainties. Finally, the composition of the total uncertainty is given Fig. 8.11 in the case of PYTHIA 8 using CSVv2; the uncertainties in each (p_T, y, flavour) bin are rescaled as follows:

$$1 = \frac{\delta^2_{\text{JEC}}}{\delta^2_{\text{TOT}}} + \frac{\delta^2_{\text{JER}}}{\delta^2_{\text{TOT}}} + \cdots \tag{8.17}$$

All uncertainties have been considered except the luminosity uncertainty (which will be added only in Chap. 9). The figure is organised in the same way as the figure for the cross sections. For n jets (b jets), the dominant uncertainty is the JEC (b calibration) in white (dark and light grey).

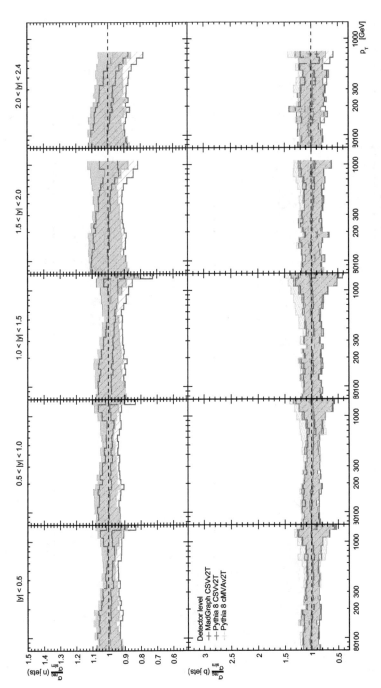

Fig. 8.8 Ratio of double differential cross section in transverse momentum and rapidity of the data with the simulations at detector level, before unfolding. The rows (columns) correspond to the flavour (rapidity). The bands corresponds to the total uncertainty

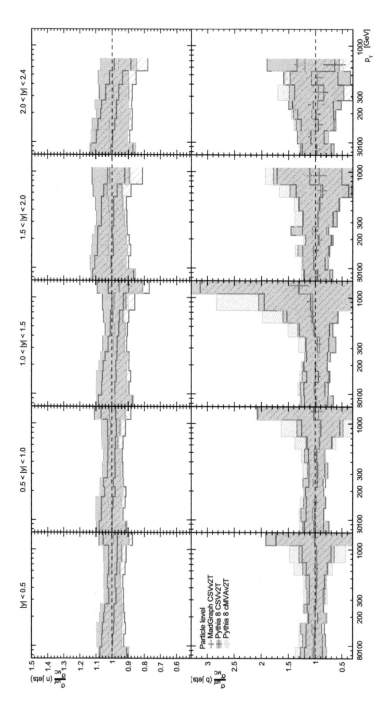

Fig. 8.9 Ratio of double differential cross section in transverse momentum and rapidity of the data with the simulation at particle level, before unfolding (the unfolding is performed with the same simulation, respectively). The rows (columns) correspond to the flavour (rapidity). The bands corresponds to the total uncertainty

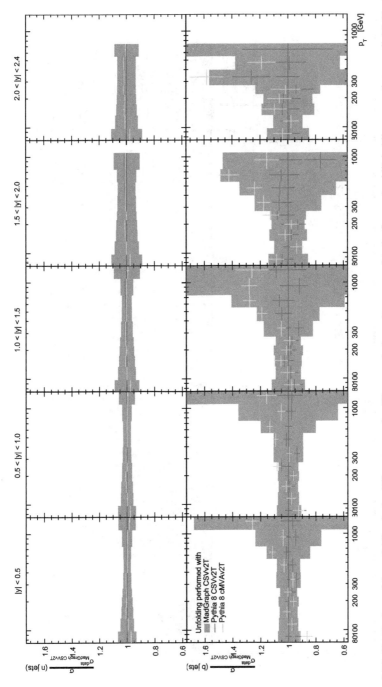

Fig. 8.10 Comparison of the unfolding in the different scenarios. The rows (columns) correspond to the flavour (rapidity). All cross sections are divided by the cross section obtained after unfolding with PYTHIA 8 (CSVv2). Systematic and statistical (onle statistical) uncertainties are shown for PYTHIA 8 using CSVv2 (for the two other scenarios)

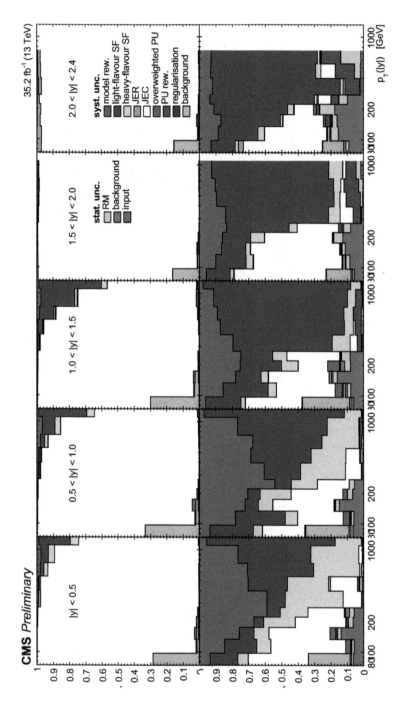

Fig. 8.11 Composition of the uncertainties after unfolding. The rows (columns) correspond to the flavour (rapidity). The uncertainties are relative, rescaled bin by bin. Both statistical and systematic uncertainties are shown

8.2.3 D'Agostini Unfolding

In principle, the unfolded results should be independent from the unfolding algorithm. In this subsection, we present an alternative method: the iterative method from D'AGOSTINI [7, 8][1] as implemented in the RooUnfold package [11].

8.2.3.1 Principles

The unfolding is formulated in terms of probabilities:

$$\hat{n}(C_i) = n(E)\mathbb{P}[C_i|E] \tag{8.18}$$

where $\hat{n}$ is the unknown distribution at particle level (C stands for *causes*) and n is the measured distribution (E stands for *effects*). The method is based on *Bayes theorem on conditional probabilities*:

$$\mathbb{P}[C_i|E_j] = \frac{\mathbb{P}[E_j|C_i]\mathbb{P}[C_i]}{\sum_{l=1}^{n_C}\mathbb{P}[E_j|C_l]\mathbb{P}[C_l]} \tag{8.19}$$

If $\mathbb{P}[C_i]$ is replaced by a MC prior (in the current case, the $p_T(y)$ spectrum at particle level in PYTHIA 8 or MADGRAPHcan be used), one can get an estimation of the true distribution. This approach of unfolding is called *Bayesian unfolding*. However, it is very biased to the MC prior; therefore, one can iterate the procedure by computing $\mathbb{P}[C_i]$ from the result of the previous iteration; with this improvement, the algorithm is called *D'Agostini's unfolding*. At each iteration, the bias to the MC prior is reduced.

Convergence. It converges to a maximum-likelihood estimator with Poisson-like errors. There cannot be any negative values, contrarily to the result from Tikhonov regularisation (which is closer to the matrix inversion). One major difficulty of the iterative approach is to determine when the convergence has been obtained, and when one should stop iterating. The number of iterations has to be determined case by case. In general, "[...] the convergence rate can be very slow and the number of iterations is expected to grow with the number of bins squared" [4].

Regularisation. The regularisation is performed by the choice of a *good* MC prior, i.e. a MC that models well the data. Therefore, the more iterations, the less biased to the MC prior, but the less regularised.

---

[1]This reference is the standard reference given in HEP. However, the same technique has already been published in other fields of science [9, 10]. In astronomy and optics for instance, it is known as *Lucy-Richardson deconvolution*.

8.2.3.2   Results

We show here that the result with D'Agostini unfolding converges and is compatible with the result with Tikhonov unfolding.

The D'Agostini unfolding is performed with different numbers of iterations: 2, 4, 8, 16 and 32. Then the different curves (various yellow levels) are then compared to the Tikhonov unfolding (blue) and the MC prior (red) in Fig. 8.12. The D'Agostini unfolding is stable after a few iterations. For small numbers of iterations, one can see that it is slightly biased to PYTHIA 8. The result after a large number of iterations is compatible within the statistical uncertainties with the Tikhonov unfolding.

Conclusions

The determination of the purity and the procedure of unfolding to particle level with determination of the uncertainties have been presented. Different scenarios involving different samples, different taggers and different unfolding algorithm lead to compatible results. Additional checks are presented in the appendix regarding the determination of the purity and the procedure of unfolding.

8.3   Details About Fit of Purity

In this appendix, we describe step by step the determination of the purity. First, we show the templates. Then we investigate the different ways to constrain charm in the CSVv2-tagged region; we compare PYTHIA 8 and MADGRAPH. After, in order to justify our approach in the non-CSVv2-tagged region, we show that the fit is not stable.

8.3.1   Templates

In the discussion conducted in Sect. 8.1, the templates were sketched in Fig. 8.2; the templates in the CSVv2-tagged (non-CSVv2-tagged) region are shown in Fig. 8.13 (Fig. 8.14) in bins of p_T and y. For readability, the statistical errors are not shown; however, they become larger and larger for increasing JP values.

Tagged region. In the CSVv2-tagged region, the different templates are peaked at different values for the different flavours; however, the peak are less and less distinct while going to higher p_T and higher y; one also observes that the c templates lie halfway between the light and b templates.

Non-tagged region. On the other hand, in the non-CSVv2-tagged region, if the templates look different at low p_T, they very similar at high p_T. Moreover, the light component has a roughly 50 times larger statistics.

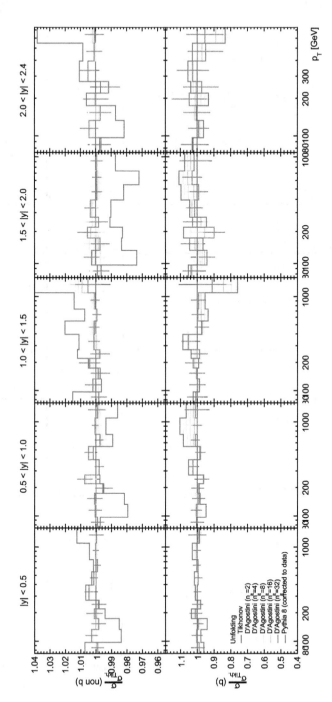

Fig. 8.12 The results of the D'Agostini unfolding with different numbers of iterations (yellow levels) are compared to the result with Tikhonov regularisation (blue, at one) and with the MC prior (red). Only the statistical uncertainty is represented. The rows (columns) correspond to the flavour (rapidity)

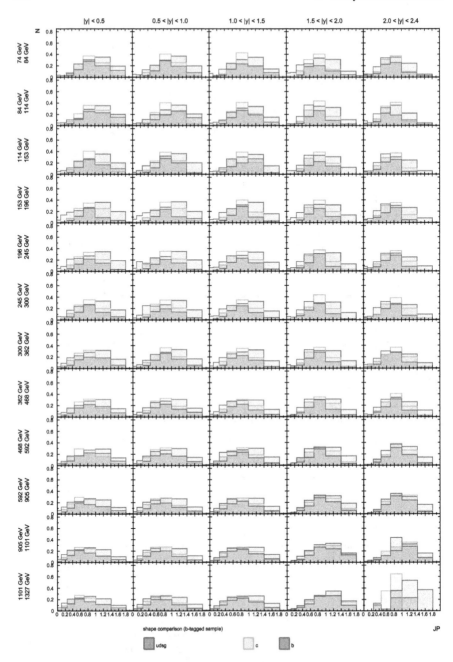

Fig. 8.13 The shape of the JP discriminant is shown for the different flavours in the CSVv2-tagged sample. Each grid corresponds to a (p_T, y) bin; the colours represent the different flavours

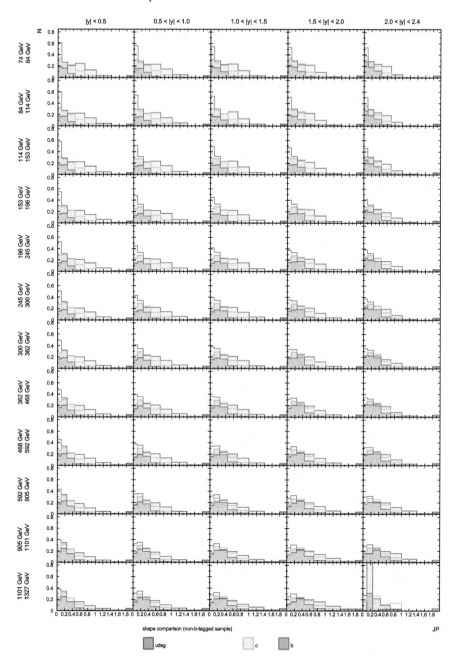

Fig. 8.14 The shape of the JP discriminant is shown for the different flavours in the non-CSVv2-tagged sample. Each grid corresponds to a (p_T, y) bin; the colours represent the different flavours

8.3.2   Determination of the Purity in the *CSVv2-Tagged Region*

In Figs. 8.15 and 8.16, the fit is investigated for different ways to constrain the c component with the PYTHIA 8 and MADGRAPHsamples:

- as an independent component, i.e. just as in Eq. 8.5 (corresponding to the blue circles);
- together with the light component, i.e. $N_{\text{data}}^{\text{total}} = p_{\text{udsg}+\text{c}}(N_{\text{MC}}^{\text{udsg}} + N_{\text{MC}}^{\text{c}}) + p_{\text{b}}N_{\text{MC}}^{\text{b}}$ (corresponding to the purple squares)
- or together with the bottom component, i.e. $N_{\text{data}}^{\text{total}} = p_{\text{udsg}}N_{\text{MC}}^{\text{udsg}} + p_{\text{b}+\text{c}}(N_{\text{MC}}^{\text{c}} + N_{\text{MC}}^{\text{b}})$ (corresponding to the orange stars).

In all cases, the configuration where charm is constrained with bottom seems optimal, since at low p_T, the correction is expected to be minimal (the disagreement in the fraction ratio mainly happens at high p_T). The findings were also confirmed using cMVAv2 (not shown here); the result was however of lower quality because the correlation of cMVAv2 with JP is greater.

The ratios of the JP discriminant in bins of p_T and y is shown before (after) the fit in Fig. 8.17 (Fig. 8.18).

8.3.3   Determination of the Purity in the Non-*CSVv2-Tagged Region*

In the non-CSVv2-tagged region, the difference of statistics of the contributions from lights and from HF components. The attempt of fit with a similar approach as in the CSVv2-tagged region (in the previous subsection) is shown in Fig. 8.19; only the case of b and c constrained together converged (in almost all bins, except at low p_T in the central region, the different attempts of fit systematically returned NaN), therefore it is the only one that can be shown here. This failure justifies the solution mentioned in Sect. 8.1, where the renormalisation factor of the $b + c$ component is propagated from the CSVv2-tagged to the non-CSVv2-tagged region, and the light component is only rescaled to match the data; the result of this procedure, with the different charm constraints, is shown in Fig. 8.20.

The ratios of the JP discriminant in bins of p_T and y is shown before (after) the fit in Fig. 8.21 (Fig. 8.22).

8.4   Details About Unfolding Procedure

We give additional details in the procedure of unfolding.

First we show the regularisation matrix obtained from PYTHIA 8. Then we discuss the treatment of the statistical uncertainties. Finally, we present additional checks.

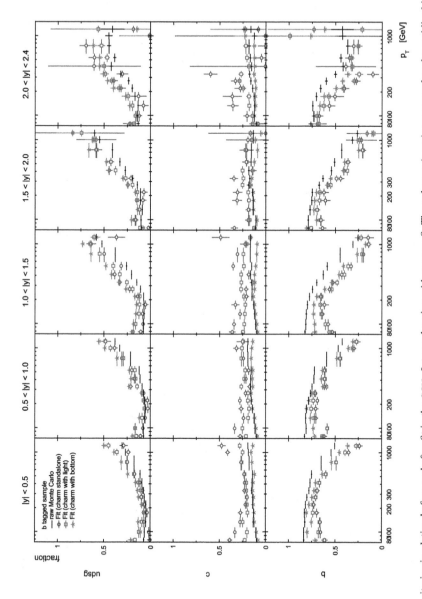

Fig. 8.15 Purity in simulation before and after fit in the CSVv2-tagged region with PYTHIA 8. The columns (rows) correspond to the rapidity bins (flavours). Different configurations to constrain charm are considered

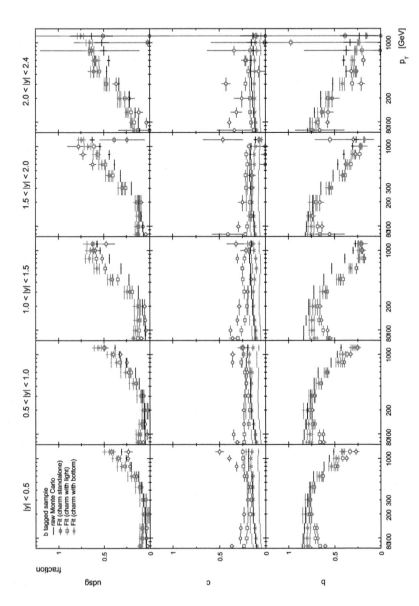

Fig. 8.16 Purity in simulation before and after fit in the CSVv2-tagged region with MADGRAPH. The columns (rows) correspond to the rapidity bins (flavours). Different configurations to constrain charm are considered

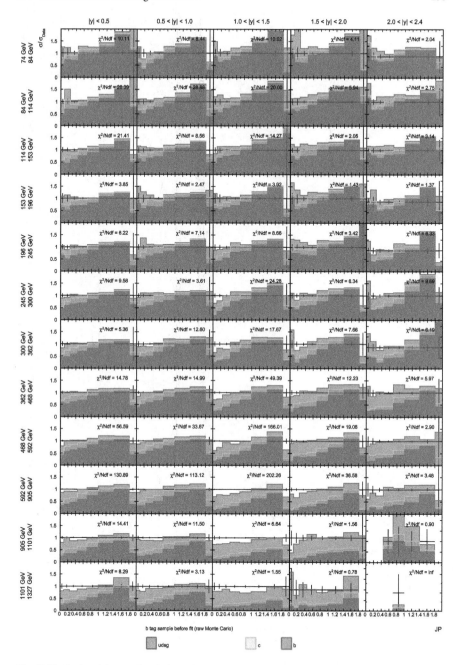

Fig. 8.17 Ratio of the JP discriminant in the CSVv2-tagged region, of simulation to data in bins of (p_T, y) before the fit; the colours stands for the flavour. The χ^2 per n.d.f. is given

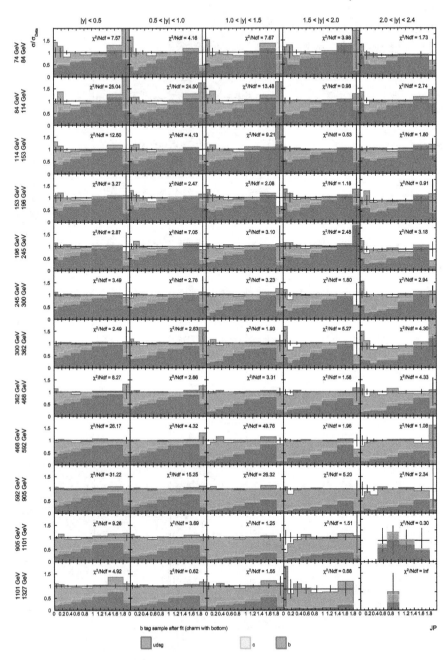

Fig. 8.18 Ratio of the JP discriminant in the CSVv2-tagged region, of simulation to data in bins of (p_T, y) after the fit with b and c constrained together; the colours stands for the flavour. The χ^2 per n.d.f. is given

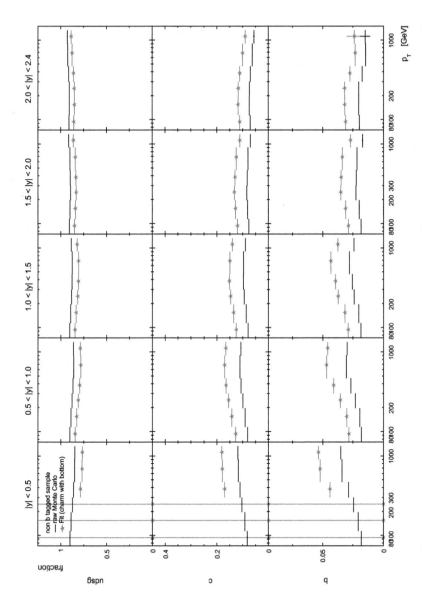

Fig. 8.19 Purity in simulation before and after fit in the CSVv2-tagged region with PYTHIA 8. The columns (rows) correspond to the rapidity bins (flavours). Only the case where charm and bottom are constrained together is considered

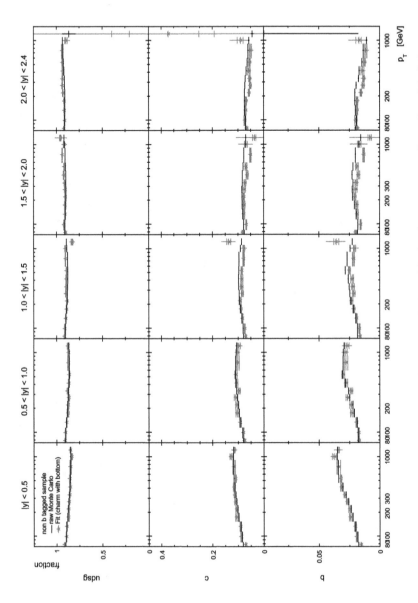

Fig. 8.20 Purity in simulation before and after extrapolation of the renormalisation of the *b* and *c* component and rescaling of the light component in the non-CSVv2-tagged region with PYTHIA 8. The columns (rows) correspond to the rapidity bins (flavours). Different configurations to constrain charm are considered

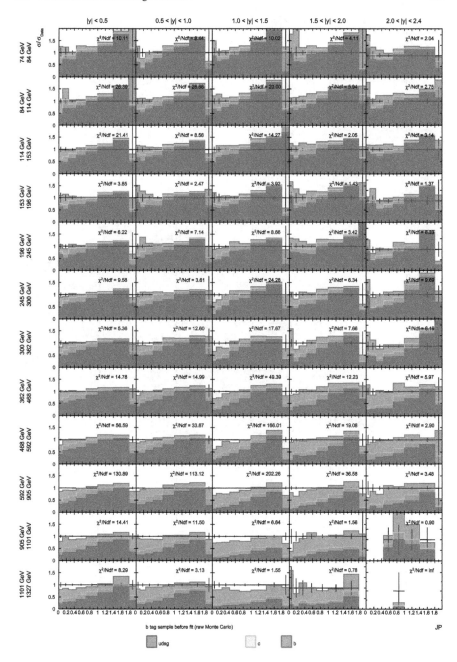

Fig. 8.21 Ratio of the JP discriminant of simulation to data in the non-CSVv2-tagged region in bins of (p_T, y) before the fit; the colours stands for the flavour. The χ^2 per n.d.f. is given

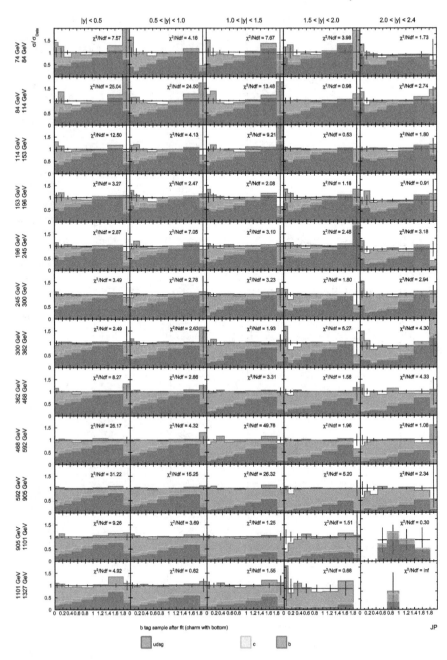

Fig. 8.22 Ratio of the JP discriminant of simulation to data in the non-CSVv2-tagged region in bins of (p_T, y) after the fit with b and c constrained together; the colours stands for the flavour. The χ^2 per n.d.f. is given

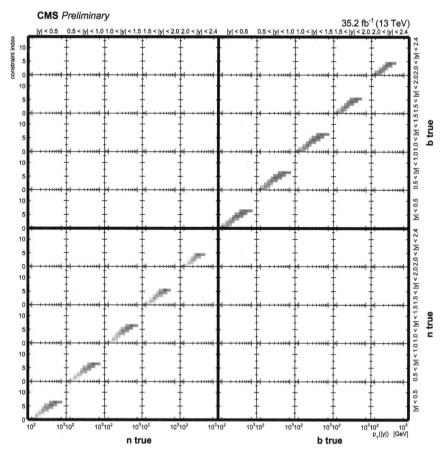

Fig. 8.23 **L** matrix constructed from PYTHIA 8. The x axis (y axis) stands for the (p_T, y, flavour) bins (constraint index). The values are given in arbitrary units; the blue (red) entries correspond to positive (negative) entries. The level of transparency is proportional to the absolute value

8.4.1 Control Plots for Tikhonov Regularisation

The **L** matrix obtained with the PYTHIA 8 sample after model reweighting may be seen on Fig. 8.23: since only the p_T is regularised, only the diagonal of the rapidity cells are filled; the constraint index corresponds to the row index in Eq. 8.14, i.e. to a constraint on three consecutive bins at particle level. The **L** matrix is therefore not a square matrix.

The effect of the regularisation on the unfolded spectra can be checked with the product **Lx** (second term in Eq. 8.11). The product is shown in the three different scenarios in Fig. 8.24. One sees explicitly which bins need the more regularisation: at high p_T, especially in the third rapidity bin.

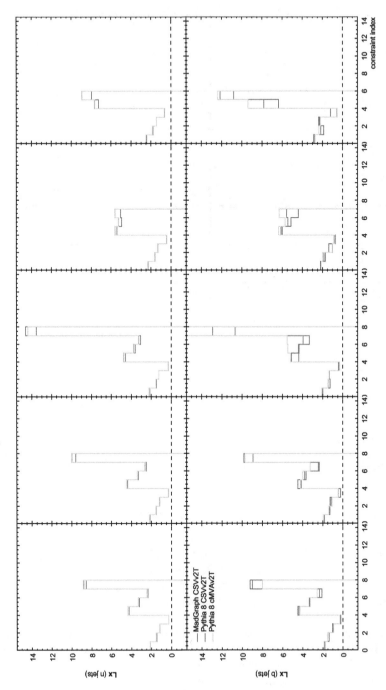

Fig. 8.24 The product of regularisation matrix and unfolded result is shown in bins of rapidity (columns) and flavour (rows), shown for the three scenarios

8.4.2   Treatment of Statistical Uncertainties

The statistical uncertainty from the MC (via the RM) and from the data (from the measurement) is considered; the former is included as an additional uncertainty in the unfolding procedure, the latter is part of the unfolded result.

The covariance matrix in data at particle level is given by the following:

$$\mathbf{V_x} = \mathbf{B^T V_y^{-1} B^T} \tag{8.20}$$

where $\mathbf{B}$, which operates the transformation, is defined as follows:

$$\mathbf{B} = \mathbf{E A^T V_y^{-1}} \tag{8.21}$$

with

$$\mathbf{E} = \left(\mathbf{A^T V_y^{-1} A} + \tau^2 \mathbf{L^T L} \right)^{-1} \tag{8.22}$$

In the case of no regularisation ($\tau = 0$), the transformation simplifies to $\mathbf{B} = \mathbf{A}^{-1}$, as expected for matrix inversion.

The covariance matrix before (after) unfolding can be seen on Fig. 8.25 (Fig. 8.26). The input covariance matrix contains only positive entries; off-diagonal events corresponds to correlations among jets coming from the same events. A single-count observable would show purely diagonal covariance matrices; here, since we are measuring a multi-count observable, there are significant non-diagonal contributions, which matter in the unfolding (see Eq. 8.10). The output covariance matrix contains negative entries (which translates into this chess-pattern); indeed, close bins are constrained together and are therefore correlated.

8.4.3   Additional Checks

We present here some additional checks to certify the unfolding. We compare systematically the result of the unfolding obtained with the D'Agostini and Tikhonov regularisations.

8.4.3.1   Backfolding

The *backfolding* consists in applying the PM on the particle-level spectrum

$$\mathbf{y'} = \mathbf{Ax} \tag{8.23}$$

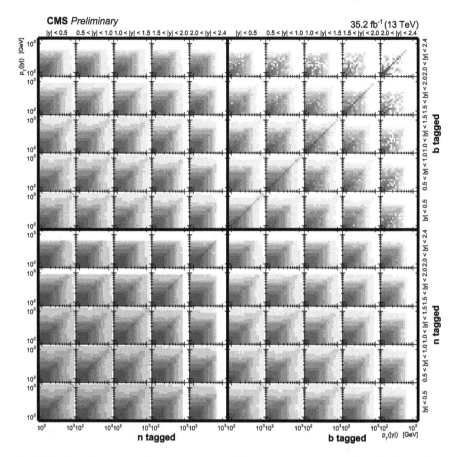

Fig. 8.25 Covariance matrix from measurement. The large sectors correspond to the flavour bins, the cells to the rapidity bins and the small matrices to the p_T bins. The level of transparency denotes the magnitude of the content in arbitrary units. All entries are positive. Off-diagonal entries show correlations among jets coming from the same events

The backfolded spectrum $\mathbf{y}'$ can be compared with the measurement $\mathbf{y}$. The difference is expected be of the order of the statistical fluctuations; however, since the backfolded spectrum still keeps track of the regularisation (either from the MC prior with D'Agostini or from the $\mathbf{L}$ matrix with Tikhonov), therefore fluctuations are expectable.

The backfolding after the two algorithms is shown in Fig. 8.27. The curves are compatible with the statistical uncertainties, both for $\hat{n}$ (above) and $\hat{b}$ jets (below). The remaining fluctuations are similar for the two backfolded spectra (with the different algorithms) and for the simulation, and give an estimate of the effect of the regularisation.

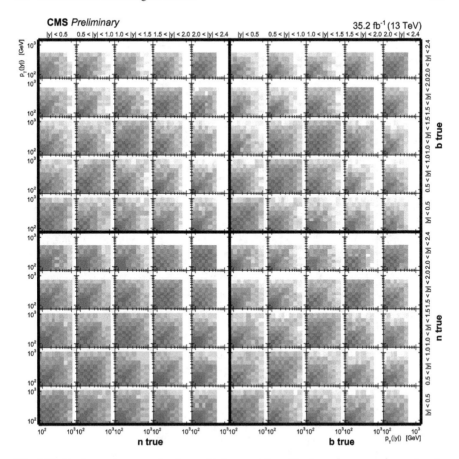

Fig. 8.26 Total covariance matrix after unfolding procedure. The large sectors correspond to the flavour bins, the cells to the rapidity bins and the small matrices to the p_T bins. The level of transparency denotes the magnitude of the content in arbitrary units. The positive (negative) entries are coloured in blue (red)

8.4.3.2 χ^2 of Agreement

The χ^2 of agreement is defined in Eq. 8.10. It is shown on the left hand side of Fig. 8.28 for the different iterations and for the unfolding obtained with Tikhonov. One observes the converge to the D'Agostini unfolding to a value close to the one of Tikhonov.

8.4.3.3 χ^2 of Change

The χ^2 of change is defined as follows:

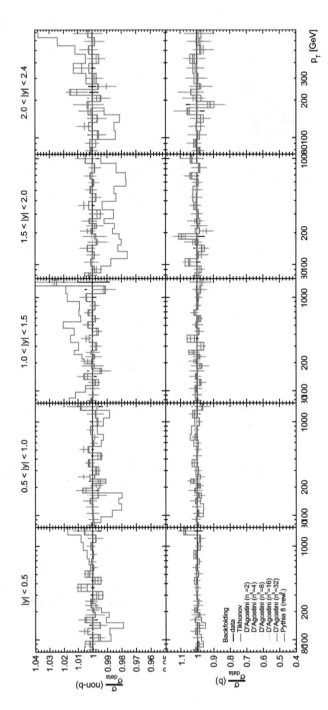

Fig. 8.27 The backfolding is compared for Tikhonov (blue) and D'Agostini (yellow) algorithms with the measurement (black, at one) and with PYTHIA 8 using CSVv2 (red). Different numbers of iterations are shown for D'Agostini. The rows (columns) correspond to the flavour (rapidity)

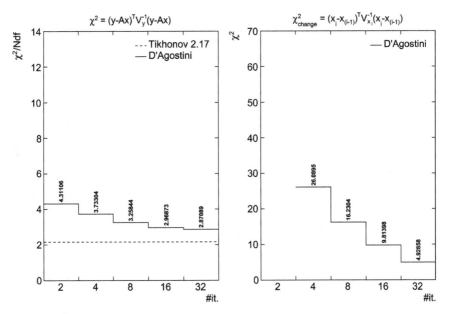

Fig. 8.28 χ^2s of agreement (left) and of change (right). The bins corresponds to different numbers of iterations in D'Agostini unfolding. The result obtained with D'Agostini (Tikhonov) unfolding is shown by a continuous (dashed) line

$$\chi^2 = (\mathbf{x}_i - \mathbf{x}_{i-1})^\mathsf{T} \mathbf{V}^{-1} (\mathbf{x}_i - \mathbf{x}_{i-1}) \tag{8.24}$$

where i denotes the iteration. It is shown on the right hand side of Fig. 8.28, where it is given for the change 2^i to 2^{i+1} iterations. The change is smaller and smaller, indicating the convergence.

8.4.3.4 Bottom Line Test

In the Bottom Line Test (BLT), we compare the agreement of simulation and data before unfolding, after unfolding and after backfolding by computing the following χ^2:

$$\chi^2_{\mathrm{BLT}} = (\mathbf{z}_{\mathrm{data}} - \mathbf{z}_{\mathrm{MC}})^\mathsf{T} \mathbf{V}^{-1}_{\mathrm{data}} (\mathbf{z}_{\mathrm{data}} - \mathbf{z}_{\mathrm{MC}}) \tag{8.25}$$

where $\mathbf{z} = \mathbf{y}$ (before) or $\mathbf{z} = \mathbf{x}$ (after) with respective data covariance matrix. One compares the values in the Tikhonov algorithm and for different number of iterations in the D'Agostini algorithm. If the unfolding is correctly performed, i.e. if only the effect of the detector is treated, then one does not expect the agreement to change significantly at the different levels.

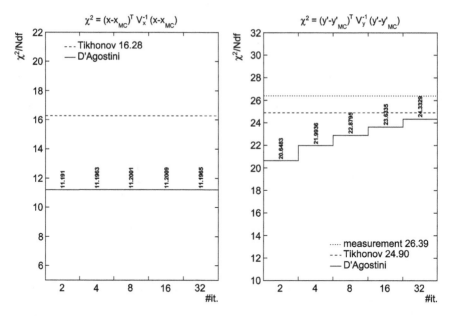

Fig. 8.29 Bottom Line Test, on the left (right) at hadron-level (detector-level) for the unfolding (backfolding). The iteration value corresponds to the one used in D'Agostini unfolding (continuous line), while the Tikhonov unfolding (dashed line) and the measurement (dashed line) has only one value

The test performed with the PYTHIA 8 sample can be seen in Fig. 8.29, where the value of Eq. 8.25 are shown for the different number of iterations of the D'Agostini unfolding in the bins and with a single line for the Tikhonov unfolding:

- The BLT of the unfolding is shown on the left hand side of the figure. The higher number of iterations does not improve the global agreement of the result obtained with the D'Agostini algorithm. Moreover, the result obtained from D'Agostini has larger uncertainties, and therefore leads to a lower χ^2_{BLT} than with the Tikhonov algorithm.

- The BLT of the backfolding is shown on the right hand side of the figure. In contrast to the BLT at particle level, the χ^2_{BLT} takes different values for D'Agostini, which is likely related to the treatment of the uncertainties. However, it goes to values of the same order as the result obtained with the Tikhonov algorithm and as the measurement. The fact the backfolding after Tikhonov algorithm has a lower χ^2_{BLT} than the measurement is explained by the regularisation; indeed, the backfolded spectrum is still regularised, with respect to the measurement.

References

1. Marchesini I, Skovpen K, Discussions on calibration with JP. Private communication
2. Kuusela M (2014) Introduction to unfolding in high energy physics. http://smat.epfl.ch/~kuusela/talks/ETH_Jul_2014.pdf. Accessed 30 Nov 2017
3. Schmitt S (2012) TUnfold: an algorithm for correcting migration effects in high energy physics. JINST 7: T10003. https://doi.org/10.1088/1748-0221/7/10/T10003, arXiv:1205.6201 [physics.data-an]
4. Stefan Schmitt. "Data Unfolding Methods in High Energy Physics". In: *EPJ Web Conf.* 137 (2017), p. 11008. https://doi.org/10.1051/epjconf/201713711008.arXiv:1611.01927 [physics.data-an]
5. Hocker A, Kartvelishvili V (1996) SVD approach to data unfolding. Nucl. Instrum. Meth. A372:469-481. https://doi.org/10.1016/0168-9002(95)01478-0. arXiv:hep-ph/9509307 [hep-ph]
6. Schmitt S (2017) Personal homepage. http://www.desy.de/~sschmitt/tunfold.html. Accessed 24 Sept 2017
7. D'Agostini G (1995) A Multidimensional unfolding method based on Bayes' theorem. Nucl Instrum Method A362:487–498. https://doi.org/10.1016/0168-9002(95)00274-X
8. D'Agostini G (2010) Improved iterative Bayesian unfolding. arXiv:1010.0632 [physics.data-an]
9. Mülthei HN, Schorr B (1987) On an iterative method for the unfolding of spectra. Nucl Instrum Methods Phys Res Sect A: Accel Spectrom Detect Assoc Equipm 257(2):371–377. ISSN: 0168-9002. https://doi.org/10.1016/0168-9002(87)90759-5
10. Shustov AE, Ulin SE (2015) Matrix of response functions for deconvolution of gamma-ray spectra. Phys Proc 74(Supplement C):399–404. Fundamental research in particle physics and cosmophysics. ISSN: 1875-3892. https://doi.org/10.1016/j.phpro.2015.09.210, http://www.sciencedirect.com/science/article/pii/S1875389215014091
11. Adye T (2011) Unfolding algorithms and tests using RooUnfold. arXiv:1105.1160. Comments: 6 pages, 5 figures, presented at PHYSTAT 2011, CERN, Geneva, Switzerland, January 2011, to be published in a CERN Yellow Report, 313-318. 6 p. https://cds.cern.ch/record/1349242

Chapter 9
Results

We compare the measurement to theory predictions. We first compare to LO predictions with PYTHIA 8, MADGRAPHand HERWIG++; then we compare to NLO predictions with POWHEG including theoretical uncertainties.

9.1 Comparison to LO Predictions

We compare the measurement with three different predictions:

- PYTHIA 8 + CUETP8M1
- MADGRAPH+ CUETP8M1
- HERWIG+++ CUETHppS1

Their specifications were given in Chap. 6.

The hadron-level inclusive jet (b jet) double differential cross section as a function of the transverse momentum and rapidity is shown in Fig. 9.1 (Fig. 9.3) and compared to the predictions in Fig. 9.2 (Fig. 9.4).

The spectrum is measured over six order of magnitude, covering a large p_T range from 74 GeV up to the TeV scale. With respect to the previous measurements at 8 and 13 TeV at CMS [1, 2], the inclusive jet cross section covers a similar p_T range, but with higher luminosity. The measurement of inclusive b jet cross section reaches much higher p_T values than the measurements at 7 TeV by CMS [3] and by ATLAS [4] and CMS [3].

In the absolute cross sections in Fig. 9.1 (Fig. 9.3), each curve corresponds to a rapidity bin. The different rapidity bins are rescaled with different factors so as not to overlap. The uncertainties are indicated with the yellow band; the statistical uncertainties are too small to appear on the figures

In the ratio in Fig. 9.2 (Fig. 9.4), the theoretical predictions are divided by the measurement. The rapidity bins are shown in successive panels. The band corresponds to the total relative uncertainty, including the statistical uncertainties, the luminosity

© Springer Nature Switzerland AG 2019

P. Connor, *Inclusive b Jet Production in Proton-Proton Collisions*, Springer Theses, https://doi.org/10.1007/978-3-030-34383-5_9

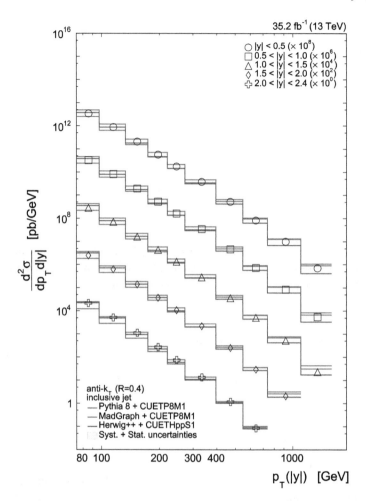

Fig. 9.1 Double differential cross section of inclusive jet production for the measurement and for LO simulation obtained with PYTHIA 8+CUETP8M1, MADGRAPH+CUETP8M1, and HER-WIG+++CUETHppS1. The markers correspond to the data; the continuous lines to the MC predictions; the yellow band to the uncertainty. The different rapidity bins have been rescaled with various factors so as no to overlap

uncertainty and the uncertainties described in Sect. 8.2.26, all added in quadrature; in addition, the statistical uncertainties are shown with vertical bars (the horizontal ticks separate the contribution from statistical errors of the measurements and the contribution inferred from MC through the unfolding). The relative uncertainty for inclusive jet (inclusive b jet) is below 10% everywhere (around $10-20\%$ up to a few hundred GeV).

We compare the result to the comparison at detector level of the inclusive jet and inclusive b tagged jet spectrum in Figs. 7.1, 7.2. For the inclusive jet production, the

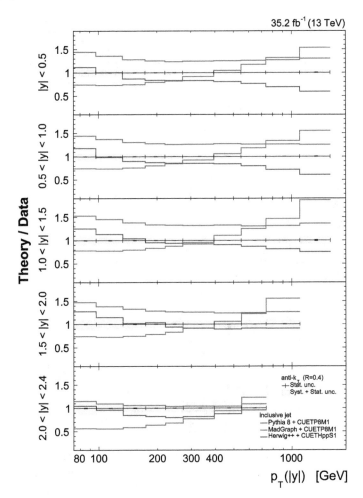

Fig. 9.2 Ratio of the double differential cross section of inclusive jet production of the measurement with LO simulation obtained with PYTHIA 8+CUETP8M1, MADGRAPH+CUETP8M1, and HERWIG+++CUETHppS1. The data is at one; the continuous lines correspond to the MC simulations and the yellow band to the total uncertainty. The statistical uncertainty is indicated with the ticks; the separation between the statistical uncertainties from the data and from the MC is indicated

ratios are similar; for the inclusive *b* jet production, we observe that the discrepancy at high-p_T has disappeared indeed, since this was due to a mis-calibration of the *b* tagging. The predictions from PYTHIA 8 and HERWIG++ are usually parallel to the measurements (except at low p_T), which may be only related to the normalisation of the simulation; on the other hand, MADGRAPH even fails at describing even the shape of the measurement.

The fraction of *b* jets is also given in Figs. 9.5, 9.6. In the fraction ratio, the simulation is divided by the measurement. The uncertainties related to jet energy and

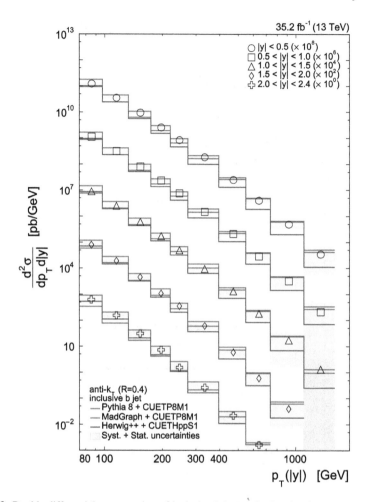

Fig. 9.3 Double differential cross section of inclusive *b* jet production for the measurement and for LO simulation obtained with PYTHIA 8+CUETP8M1, MADGRAPH+CUETP8M1, and HER-WIG+++CUETHppS1. The markers correspond to the data; the continuous lines to the MC predictions; the yellow band to the uncertainty. The different rapidity bins have been rescaled with various factors so as no to overlap

to pile-up cancel in the fraction; however, the dominant uncertainties are related to the *b* tagging calibration and to the fit of the purity, and are not vanishing in the fraction. One observes a significant difference among, on one hand, PYTHIA 8 and MADGRAPH, and, on the other hand, HERWIG++, especially for central rapidity and at high transverse momentum. The predictions by PYTHIA 8 and MAD-GRAPH shows that for $p_T \gg m_b$, the dynamics for *b* jets do not differ from the dynamics for jets.

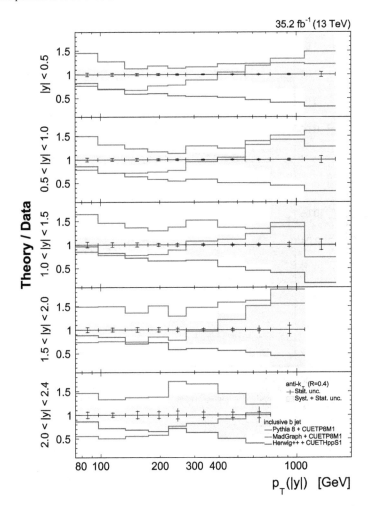

Fig. 9.4 Ratio of the double differential cross section of inclusive *b* jet production of the measurement with LO simulation obtained with PYTHIA 8+CUETP8M1, MADGRAPH+CUETP8M1, and HERWIG+++CUETHppS1. The data is at one; the continuous lines correspond to the MC simulations and the yellow band to the total uncertainty. The statistical uncertainty is indicated with the ticks; the separation between the statistical uncertainties from the data and from the MC is indicated

9.2 Comparison to NLO Predictions

We compare the measurement with POWHEG matched to CUETP8M1 [5–7].

First, we describe the simulation and the theoretical uncertainties. Second, we compare data and predictions. Finally, we investigate the different contributions in the simulation.

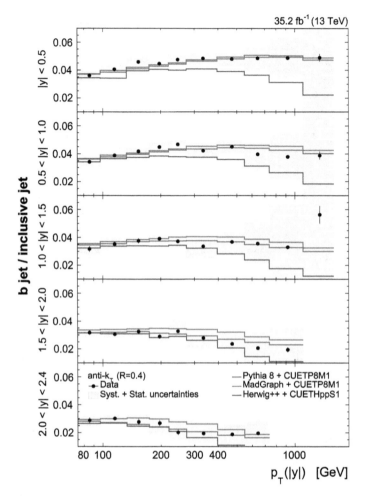

Fig. 9.5 Fraction of *b* jet production in inclusive jet production in the measurement and of LO simulation obtained with PYTHIA 8+CUETP8M1, MADGRAPH+CUETP8M1, and HER-WIG+++CUETHppS1. The markers correspond to the data; the continuous lines to the MC predictions; the yellow band to the uncertainty. The different rapidity bins have been rescaled with various factors so as no to overlap

9.2.1 Theoretical Predictions

POWHEGwas already presented in Sect. 4.2. It allows to compute predictions at NLO, include a Sudakov factor with splitting functions at NLO and is interfaced with PYTHIA 8 for the PS, the MPI and the hadronisation.

The nominal value and the uncertainties of the theory predictions are derived as follows:

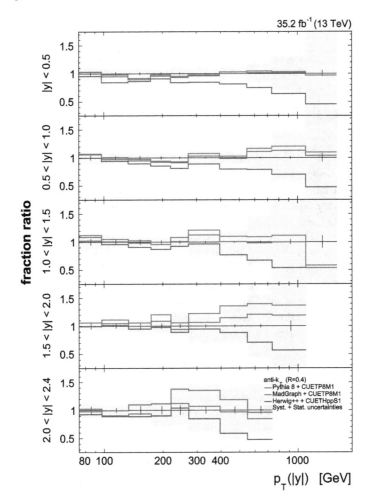

Fig. 9.6 Fraction ratio of the *b* jet production in inclusive jet production in the measurement with LO simulation obtained with PYTHIA 8+CUETP8M1, MADGRAPH+CUETP8M1, and HER-WIG+++CUETHppS1. The data is at one; the continuous lines correspond to the MC simulations and the yellow band to the total uncertainty. The statistical uncertainty is indicated with the ticks

PDF The NNPDF 3.0 set contains a hundred of replicas. In the generation process, each event is varied according to the replicas; each replica therefore leads to a different cross section. The determination of the nominal value and of the uncertainties is performed bin after bin:

- The nominal value of the cross section is taken as the mean of the variations.
- The replicas are ordered by increasing value, and the 16th (84th) variations are taken as lower (upper) variation, corresponding to a variation of $\pm 1\sigma$ up and down.

Table 9.1 Variations of the tune

	Variable	Nominal	Up	Down
MPI	Cut-off p_{T0}^{ref}	2.4024	1.8238	2.60468
	Exponent ε	0.25208	2.5208	0.25208
	Reference energy $\sqrt{s_0}$	1.6	3.2	1.5
Colour reconnection	Free parameter R	1.80	7.60	4.20

Scale variations   The renormalisation and factorisation scales are varied with a factor $1/2$ down and 2 up. Four variations are considered, corresponding to independent variations of the two scales. The envelope of the scale variations is taken as uncertainty. The joint variation does not change the uncertainty band significantly (not shown here).

PS variations   The scale in the PS is also varied with a factor $1/2$ down and 2 up.

Tune variations   The parameters of CUETP8M1 are varied, correspondingly to the values shown in Table 9.1. Not all parameters are varied, but only the ones proper to CUETP8M1, i.e. parameters related to MPI (described in Eq. 2.60) and to colour reconnection. The whole list of tune parameters can be found in Appendix 9.4.

Fragmentation functions   Finally, since the b jets are defined at hadron level, an uncertainty on the Bowler factor in the FF (Eq. 2.64) is considered:

$$r_B = 0.895^{+0.184}_{-0.197} \qquad (9.1)$$

The uncertainties are then summed in quadrature separately up and down.

Figures 9.7, 9.8 show the composition of the theoretical uncertainties normalised to unity in each bin for the inclusive jet and inclusive b jet productions. The scale uncertainties dominate the uncertainties (lower row); the PS uncertainties get larger and larger at higher and higher p_T; the fragmentation and tune uncertainties (yellow and green) are of the same order as statistical uncertainties (upper row).

9.2.2   Comparison

We show the comparison of POWHEG+PYTHIA 8 (continuous cyan curve) with theory uncertainties (dashed cyan curves) to the measurement in Figs. 9.9, 9.10, 9.11, 9.12, 9.13 and 9.14 for the inclusive jet and inclusive b jet productions.

The figures are organised in the same way as in the first section: first, the inclusive jet cross section is presented in Figs. 9.9, 9.10. The agreement of data and simulation is excellent; the data has smaller uncertainty bands than the predictions. Then, the inclusive b jet cross section is presented in Figs. 9.11, 9.12. Data and simulation also agree within the uncertainties; fluctuations in data are likely due to the irregularities in the performance of the taggers.

The comparison to the fraction in the inclusive jet production is shown in Figs. 9.13, 9.14. The theory agrees everywhere within the theoretical and experi-

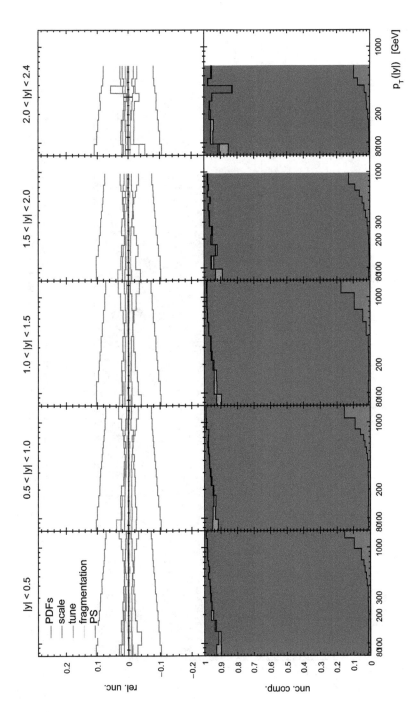

Fig. 9.7 Composition of the theory uncertainties for the POWHEG+ CUETP8M1 prediction of the inclusive jet production. The uncertainties are stacked and the sum is normalised to unity, bin after bin. Above (below) is the relative uncertainties (is the stack of the relative uncertainties normalised to one bin by bin). The five bins of rapidity are shown in columns

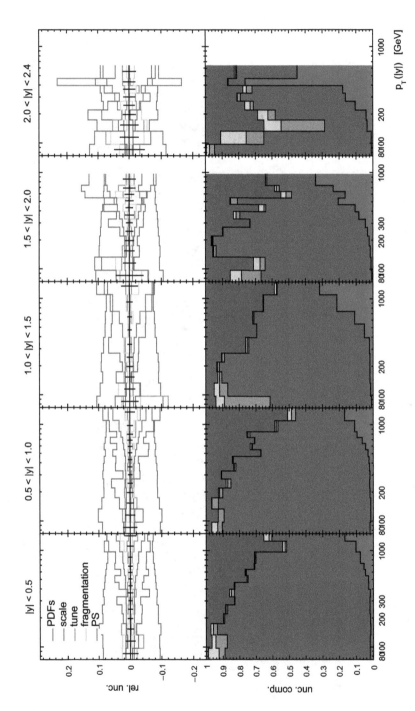

Fig. 9.8 Composition of the theory uncertainties for the POWHEG+ CUETP8M1 prediction of the inclusive *b* jet production. The uncertainties are stacked and the sum is normalised to unity, bin after bin. Above (below) is the relative uncertainties (is the stack of the relative uncertainties normalised to one bin by bin). The five bins of rapidity are shown in columns

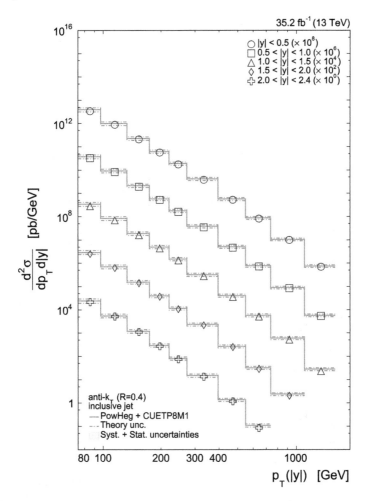

Fig. 9.9 Double differential cross section of inclusive jet production for the measurement and for NLO simulation obtained with POWHEG+CUETP8M1. The markers correspond to the data; the continuous lines to the MC predictions; the yellow band to the uncertainty. The different rapidity bins have been rescaled with various factors so as no to overlap

mental uncertainties; however, in this case, the theory uncertainties are smaller than the experimental uncertainties.

9.2.3 Contributions to the Prediction

In Figs. 9.15, 9.16, we investigate the different contributions to the predictions:

– hard process + PS + MPI + hadronisation

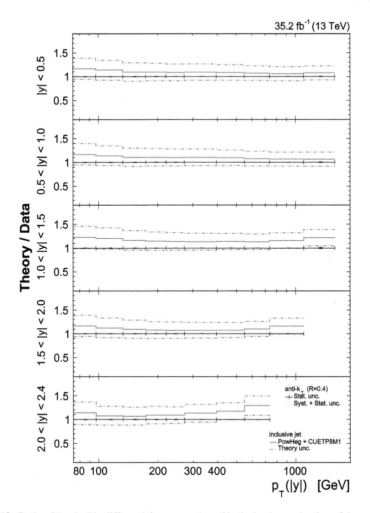

Fig. 9.10 Ratio of the double differential cross section of inclusive jet production of the measurement with NLO simulation obtained with POWHEG+CUETP8M1. The data is at one; the continuous lines correspond to the MC simulations and the yellow band to the total uncertainty. The statistical uncertainty is indicated with the ticks; the separation between the statistical uncertainties from the data and from the MC is indicated

– hard process + PS + hadronisation
– hard process + hadronisation

The two figures correspond to the cases of the inclusive jet and inclusive b jet productions. The uncertainties correspond to the scale and PDF uncertainties (described in the next section).

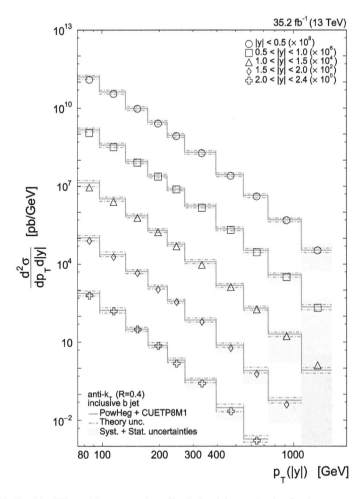

Fig. 9.11 Double differential cross section of inclusive *b* jet production for the measurement and for NLO simulation obtained with POWHEG+CUETP8M1. The markers correspond to the data; the continuous lines to the MC predictions; the yellow band to the uncertainty. The different rapidity bins have been rescaled with various factors so as no to overlap

From the figures, one concludes that MPI has mainly an effect at low p_T (from red to blue), both for jets and *b* jets. However, the PS has a different effect for jets and for *b* jets. The PS "unsmears" the spectrum, since high p_T partons are likely to radiate; however, the *b* jet spectrum is significantly reduced, since a significant fraction of the *b* quarks come from gluon splitting. This effect was already investigated in Chap. 6 (Figs. 6.1, 6.2).

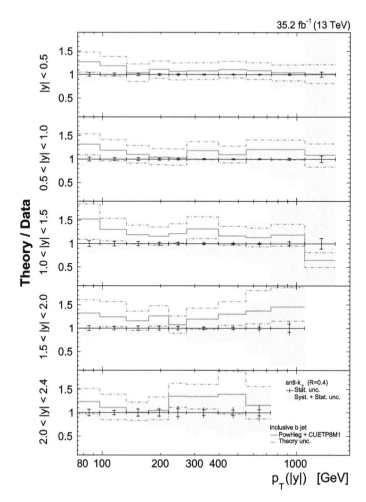

Fig. 9.12 Ratio of the double differential cross section of inclusive*b* jet production of the measurement with NLO simulation obtained with POWHEG+CUETP8M1. The data is at one; the continuous lines correspond to the MC simulations and the yellow band to the total uncertainty. The statistical uncertainty is indicated with the ticks; the separation between the statistical uncertainties from the data and from the MC is indicated

9.3 Tables

The systematic uncertainties are given for each source in Tables 9.2, 9.3 and 9.4. It can be seen that at high p_T the very dominant uncertainty comes from the flavour-related calibration, namely the two uncertainties for light- and heavy-flavour scale factors, as well as the model reweighting described in the main text.

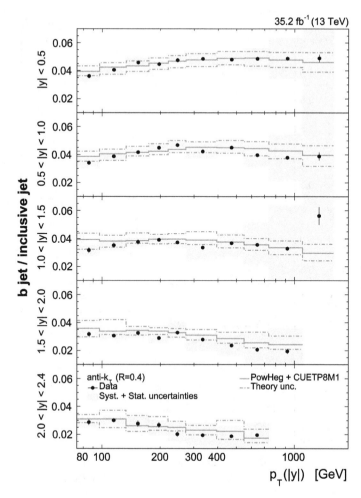

Fig. 9.13 Fraction of the b jet production in the inclusive jet production in the measurement and of NLO simulation obtained with POWHEG+CUETP8M1. The markers correspond to the data; the continuous lines to the MC predictions; the yellow band to the uncertainty. The different rapidity bins have been rescaled with various factors so as no to overlap

9.4 List of Tune Parameters

CUETP8M1 is a tune developed by and for CMS, based on the Monash 2013 tune for PYTHIA 8.1 [8].

The Monash tune gathers the parameters from e^+e^- measurements at LEP and SLD[1] (especially all parameters related to hadronisation and FSR), and

---

[1] SLAC Large Detector.

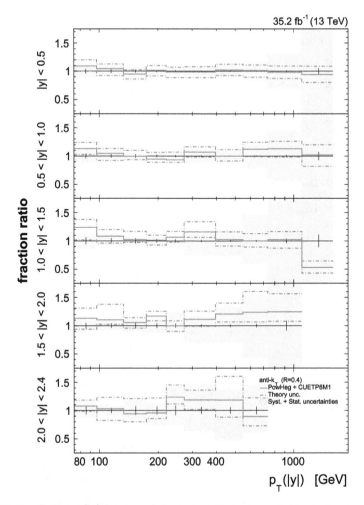

Fig. 9.14 Fraction ratio of the *b* jet production in the inclusive jet production in the measurement with NLO simulation obtained with POWHEG+CUETP8M1. The data is at one; the continuous lines correspond to the MC simulations and the yellow band to the total uncertainty. The statistical uncertainty is indicated with the ticks

parameters from *pp* and *p p̄* measurements at TEVATRON and LHC (FSR, MPI, etc.).

CUETP8M1 essentially contains a retuning of MPI, which is expected to be more significant at higher energy in the centre-of-mass.

All parameters, as used in the POWHEG prediction, are shown in Table 9.5.

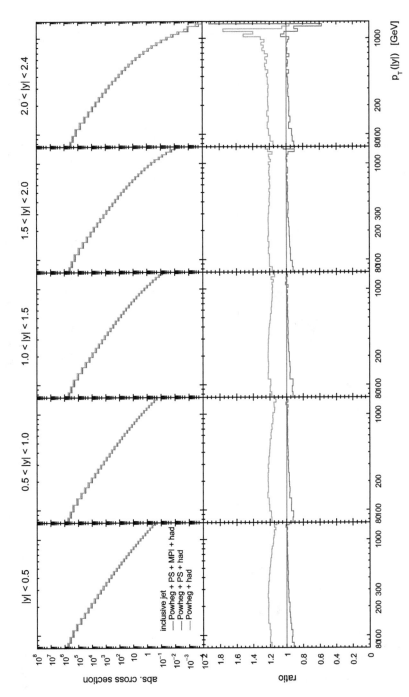

Fig. 9.15 Contributions to the prediction of the inclusive jet production by POWHEG+ CUETP8M1. The columns correspond to the rapidity bins. Above (below) is the absolute cross section as predicted by POWHEG+ CUETP8M1 (the ratio to the prediction including all contributions)

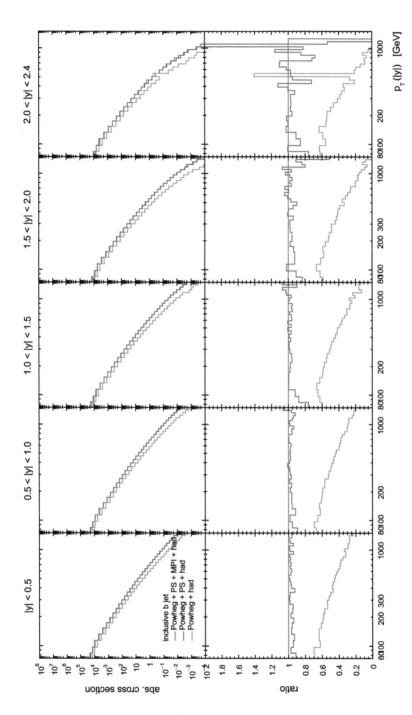

Fig. 9.16 Contributions to the prediction of the inclusive *b* jet production by POWHEG+ CUETP8M1. The columns correspond to the rapidity bins. Above (below) is the absolute cross section as predicted by POWHEG+ CUETP8M1 (the ratio to the prediction including all contributions)

Table 9.2 Inclusive jet measurement and systematic uncertainties

y	p_T(GeV)	Cross section	Tot. syst. unc. (%)	Background (%)	Regularisation (%)	PU reweighting (%)	Overweighted PU (%)	JEC (%)	JER (%)	Lumi (%)
$\lvert y\rvert < 0.5$	74–97	7.8×10^5	+8.1 / −7.2	+4.0 / −3.9	+0.0 / −0.0	+0.3 / −0.3	+3.3 / −0.1	+5.8 / −5.5	+0.2 / −0.2	+2.5 / −2.5
	97–133	3.1×10^5	+6 / −5.6	+0.5 / −0.5	+0.0 / −0.0	+0.2 / −0.1	+2.1 / −0.1	+4.9 / −4.5	+0.2 / −0.1	+2.5 / −2.5
	133–174	8.5×10^4	+4.8 / −5.2	+0.4 / −0.4	+0.0 / −0.0	+0.0 / −0.0	+0.4 / −1.6	+4.1 / −4.3	+0.1 / −0.1	+2.5 / −2.5
	174–220	2.6×10^4	+4.4 / −4.4	+0.0 / −0.0	+0.0 / −0.0	+0.0 / −0.0	+0.3 / −0.7	+3.7 / −3.5	+0.1 / −0.1	+2.5 / −2.5
	220–272	9.1×10^3	+4.3 / −4.2	+0.0 / −0.0	+0.0 / −0.0	+0.0 / −0.0	+0.2 / −0.3	+3.5 / −3.4	+0.1 / −0.1	+2.5 / −2.5
	272–395	4.7×10^3	+3.9 / −3.8	+0.0 / −0.0	+0.0 / −0.0	+0.1 / −0.1	+0.1 / −0.1	+2.9 / −2.8	+0.2 / −0.0	+2.5 / −2.5
	395–548	8×10^2	+3.6 / −3.5	+0.0 / −0.0	+0.0 / −0.0	+0.0 / −0.2	+0.0 / −0.0	+2.6 / −2.5	+0.1 / −0.0	+2.5 / −2.5
	548–737	1.5×10^2	+3.6 / −3.6	+0.0 / −0.0	+0.0 / −0.0	+0.0 / −0.1	+0.0 / −0.0	+2.6 / −2.6	+0.0 / −0.1	+2.5 / −2.5
	737–1101	36	+4.3 / −4.2	+0.0 / −0.0	+0.0 / −0.0	+0.0 / −0.0	+0.1 / −0.0	+3.3 / −3.3	+0.1 / −0.1	+2.5 / −2.5
	1101–1588	3.4	+5.9 / −5.7	+0.1 / −0.1	+0.0 / −0.0	+0.0 / −0.0	+0.1 / −0.1	+5.4 / −5.2	+0.1 / −0.1	+2.5 / −2.5
$0.5 < \lvert y\rvert < 1.0$	74–97	7.5×10^5	+7.8 / −7	+4.1 / −4.0	+0.0 / −0.0	+0.3 / −0.3	+3.1 / −0.1	+5.3 / −5.2	+0.4 / −0.3	+2.5 / −2.5
	97–133	2.9×10^5	+5.5 / −5.5	+0.5 / −0.5	+0.0 / −0.0	+0.1 / −0.1	+2.0 / −2.2	+4.4 / −4.3	+0.2 / −0.3	+2.5 / −2.5
	133–174	7.9×10^4	+5 / −4.9	+0.4 / −0.4	+0.0 / −0.0	+0.1 / −0.0	+0.5 / −1.5	+4.2 / −3.9	+0.2 / −0.1	+2.5 / −2.5
	174–220	2.4×10^4	+4.2 / −4.4	+0.0 / −0.0	+0.0 / −0.0	+0.0 / −0.2	+0.3 / −0.4	+3.3 / −3.6	+0.2 / −0.3	+2.5 / −2.5
	220–272	8.3×10^3	+4.4 / −4.3	+0.0 / −0.0	+0.0 / −0.0	+0.1 / −0.1	+0.1 / −0.1	+3.6 / −3.4	+0.2 / −0.1	+2.5 / −2.5
	272–395	4.3×10^3	+3.9 / −3.9	+0.0 / −0.0	+0.0 / −0.0	+0.1 / −0.1	+0.3 / −0.0	+3.0 / −2.9	+0.1 / −0.2	+2.5 / −2.5
	395–548	7.1×10^2	+3.8 / −3.7	+0.0 / −0.0	+0.0 / −0.0	+0.0 / −0.0	+0.0 / −0.0	+2.8 / −2.7	+0.1 / −0.1	+2.5 / −2.5
	548–737	1.4×10^2	+3.7 / −3.6	+0.0 / −0.0	+0.0 / −0.0	+0.1 / −0.1	+0.1 / −0.0	+2.7 / −2.6	+0.1 / −0.1	+2.5 / −2.5
	737–1101	31	+4 / −3.9	+0.0 / −0.0	+0.0 / −0.0	+0.0 / −0.0	+0.3 / −0.0	+3.1 / −3.0	+0.2 / −0.1	+2.5 / −2.5
	1101–1588	2.6	+5.3 / −5.1	+0.1 / −0.1	+0.0 / −0.0	+0.0 / −0.0	+0.1 / −0.0	+4.7 / −4.5	+0.1 / −0.1	+2.5 / −2.5
$1.0 < \lvert y\rvert < 1.5$	74–97	6.5×10^5	+10 / −9.1	+5.3 / −5.2	+0.0 / −0.0	+0.4 / −0.4	+3.4 / −0.3	+7.6 / −7.1	+0.5 / −0.6	+2.5 / −2.5
	97–133	2.5×10^5	+7.5 / −7.1	+0.9 / −1.0	+0.0 / −0.0	+0.1 / −0.1	+2.6 / −2.7	+6.5 / −6.0	+0.6 / −0.5	+2.5 / −2.5
	133–174	6.6×10^4	+6.2 / −6.2	+0.7 / −0.7	+0.0 / −0.0	+0.0 / −0.0	+0.3 / −1.5	+5.6 / −5.4	+0.1 / −0.5	+2.5 / −2.5
	174–220	2×10^4	+6 / −5.5	+0.1 / −0.1	+0.0 / −0.0	+0.3 / −0.1	+0.6 / −0.7	+5.4 / −4.9	+0.5 / −0.1	+2.5 / −2.5

(continued)

Table 9.2 (continued)

y	p_T(GeV)	Cross section	Tot. syst. unc. (%)	Background (%)	Regularisation (%)	PU reweighting (%)	Overweighted PU (%)	JEC (%)	JER (%)	Lumi (%)
	220–272	6.8×10^3	+5.6 / −5.4	+0.1 / −0.1	+0.0 / −0.0	+0.1 / −0.2	+0.2 / −0.2	+5.0 / −4.8	+0.3 / −0.4	+2.5 / −2.5
	272–395	3.5×10^3	+4.8 / −4.8	+0.0 / −0.0	+0.0 / −0.0	+0.1 / −0.0	+0.1 / −0.2	+4.1 / −4.1	+0.2 / −0.2	+2.5 / −2.5
	395–548	5.5×10^2	+4.5 / −4.2	+0.0 / −0.0	+0.0 / −0.0	−0.0 / −0.1	−0.2 / −0.0	+3.7 / −3.4	+0.4 / −0.2	+2.5 / −2.5
	548–737	97	+4 / −3.9	+0.0 / −0.0	+0.0 / −0.0	−0.1 / −0.0	+0.1 / +0.1	+3.1 / −3.0	+0.2 / +0.3	+2.5 / −2.5
	737–1101	19	+4.6 / −4.5	−0.0 / −0.0	+0.0 / −0.0	−0.0 / −0.0	+0.1 / −0.0	+3.8 / −3.7	+0.3 / −0.2	+2.5 / −2.5
	1101–1588	1.1	+7.5 / −6.9	+0.1 / −0.1	+0.0 / −0.0	+0.0 / −0.0	+0.1 / −0.2	+7.0 / −6.4	+0.5 / −0.4	+2.5 / −2.5
$1.5 < |y| < 2.0$	74–97	5.8×10^5	+13 / −11	+4.9 / −4.8	+0.0 / −0.0	+0.4 / −0.4	+3.5 / −0.2	+11.1 / −9.6	+1.1 / −1.1	+2.5 / −2.5
	97–133	2.2×10^5	+9.1 / −8.8	+0.7 / −0.8	+0.0 / −0.0	+0.2 / −0.1	+2.1 / −2.5	+8.5 / −8.0	+0.5 / −0.5	+2.5 / −2.5
	133–174	5.7×10^4	+8.8 / −8	+0.6 / −0.5	+0.0 / −0.0	+0.0 / −0.1	+0.7 / −1.5	+8.4 / −7.4	+0.5 / −0.8	+2.5 / −2.5
	174–220	1.7×10^4	+8.3 / −7.6	+0.1 / −0.1	+0.0 / −0.0	+0.1 / −0.1	+0.1 / −0.5	+7.9 / −7.1	+0.5 / −0.4	+2.5 / −2.5
	220–272	5.6×10^3	+7.7 / −7.2	−0.1 / +0.0	+0.0 / −0.0	−0.1 / −0.1	+0.2 / −0.2	+7.3 / −6.8	+0.5 / −0.3	+2.5 / −2.5
	272–395	2.7×10^3	+7.4 / −6.8	+0.0 / −0.0	+0.0 / −0.0	+0.1 / −0.1	+0.0 / −0.1	+6.9 / −6.3	+0.5 / −0.2	+2.5 / −2.5
	395–548	3.8×10^2	+7.1 / −6.7	+0.0 / −0.0	−0.0 / −0.0	+0.1 / −0.1	+0.1 / −0.0	+6.7 / −6.2	+0.2 / −0.4	+2.5 / −2.5
	548–737	55	+7.5 / −7.1	+0.0 / −0.0	+0.0 / −0.0	+0.2 / −0.2	+0.0 / −0.1	+7.1 / −6.6	+0.1 / −0.5	+2.5 / −2.5
	737–1101	7.5	+10 / −9.1	−0.0 / −0.1	−0.0 / −0.0	+0.1 / −0.0	−0.1 / −0.1	+9.7 / −8.7	+0.5 / −0.4	+2.5 / −2.5
$2.0 < |y| < 2.4$	74–97	4×10^5	+13 / −11	+4.8 / −4.7	+0.0 / −0.0	+0.7 / −0.7	+3.7 / −0.1	+10.8 / −9.5	+1.9 / −1.8	+2.5 / −2.5
	97–133	1.5×10^5	+8.7 / −8.4	+0.5 / −0.5	+0.0 / −0.0	+0.5 / −0.4	+1.7 / −2.3	+8.0 / −7.6	+1.3 / −1.2	+2.5 / −2.5
	133–174	3.6×10^4	+8.4 / −7.6	+0.6 / −0.6	+0.0 / −0.0	+0.1 / −0.1	+0.6 / −1.1	+7.9 / −7.0	+0.8 / −1.0	+2.5 / −2.5
	174–220	1×10^4	+7.6 / −7	−0.1 / −0.1	+0.0 / −0.0	−0.3 / +0.4	−1.1 / +0.4	+7.1 / −6.4	+1.0 / −0.7	+2.5 / −2.5
	220–272	3.1×10^3	+7.4 / −7.1	+0.1 / −0.1	+0.0 / −0.0	+0.1 / −0.0	−0.7 / +0.1	+6.9 / −6.6	+0.5 / −1.0	+2.5 / −2.5
	272–395	1.3×10^3	+7.1 / −6.8	+0.1 / −0.1	+0.0 / −0.0	−0.2 / +0.1	+0.1 / −0.1	+6.6 / −6.3	+0.8 / −0.7	+2.5 / −2.5
	395–548	1.4×10^2	+7.4 / −7.1	+0.0 / −0.0	−0.0 / −0.0	−0.0 / −0.1	+0.1 / −0.0	+7.0 / −6.6	+0.8 / −1.0	+2.5 / −2.5
	548–737	13	+9.5 / −8.5	+0.3 / −0.3	+0.0 / −0.0	+0.1 / −0.1	+0.1 / −0.0	+9.1 / −8.1	+1.1 / −1.0	+2.5 / −2.5

Table 9.3 Inclusive b jet measurement and systematic uncertainties

y	p_T(GeV)	Cross section	Tot. syst. unc. (%)	Back-ground (%)	Regulari-sation (%)	PU re-weighting (%)	Over-weighted PU (%)	JEC (%)	JER (%)	Heavy-flavour SF (%)	Light-flavour SF (%)	Model reweighting (%)	Lumi (%)
$|y| < 0.5$	74–97	2.8×10^4	+14 / −11	+5.8 / −5.9	+0.0 / −0.0	+0.8 / −0.7	+10.2 / −1.6	+2.3 / −5.3	+0.5 / −0.1	+1.7 / −1.6	+5.8 / −2.5	+3.5 / −3.7	+2.5 / −2.5
	97–133	1.3×10^4	+9.7 / −9.4	+0.7 / −0.6	+0.0 / −0.0	+0.0 / −0.0	+4.6 / −6.6	−3.6 / −6.6	+0.0 / −0.2	+2.0 / −2.1	+3.4 / −3.3	+2.7 / −2.9	+2.5 / −2.5
	133–174	3.9×10^3	+5 / −10	+0.6 / −0.5	+0.0 / −0.0	+0.4 / −0.2	+0.5 / −7.4	+1.7 / −5.2	+0.3 / +0.3	+1.9 / −3.6	+2.3 / −2.3	+2.5 / −2.8	+2.5 / −2.5
	174–220	1.2×10^3	+8 / −5.8	+0.1 / −0.1	+0.0 / −0.0	+0.7 / −0.0	+0.7 / −0.3	−5.5 / −2.5	+0.5 / −0.2	+3.6 / −2.7	+1.7 / +1.9	+2.8 / −3.2	+2.5 / −2.5
	220–272	4.3×10^2	+5.3 / −8.7	+0.0 / −0.0	+0.0 / −0.0	+1.2 / −1.4	+0.0 / −0.1	+1.2 / −3.9	+1.1 / +0.3	−2.7 / −3.8	−1.7 / −4.5	−3.5 / −3.9	+2.5 / −2.5
	272–395	2.3×10^2	+10 / −9.9	+0.0 / −0.0	+0.0 / −0.0	+0.9 / −0.4	+0.1 / −0.0	+4.2 / −3.0	+0.3 / −0.1	+5.6 / −3.8	+4.9 / −4.6	−5.4 / −6.1	+2.5 / −2.5
	395–548	38	+11 / −10	+0.0 / −0.0	+0.0 / −0.0	+0.8 / −0.0	+0.0 / −0.0	+2.5 / −2.4	+0.5 / −0.2	+5.1 / −4.9	+7.2 / −4.6	+5.3 / −6.2	+2.5 / −2.5
	548–737	7.5	+16 / −17	+0.0 / −0.0	+0.0 / −0.0	+0.6 / −0.8	+0.0 / −0.0	+1.2 / −2.2	−0.1 / −0.2	−4.0 / +8.8	−6.4 / +11.0	+7.8 / −9.2	+2.5 / −2.5
	737–1101	1.8	+30 / −30	+0.0 / −0.0	+0.0 / −0.0	+1.6 / −1.6	+0.1 / −0.1	+4.0 / −2.3	+0.2 / −0.1	+14.4 / −12.8	+22.6 / −23.0	+12.1 / −14.2	+2.5 / −2.5
	1101–1588	0.17	+83 / −86	+0.2 / −0.2	+0.0 / −0.0	+3.1 / −3.2	+0.2 / −0.1	+3.1 / −4.5	+0.1 / −0.2	+32.1 / −25.5	+75.2 / −79.9	+15.5 / −18.0	+2.5 / −2.5
$0.5 < |y| < 1.0$	74–97	2.6×10^4	+14 / −11	+5.8 / −5.9	+0.0 / −0.0	+0.7 / −0.8	+9.4 / −0.1	+3.9 / −5.0	+0.1 / −0.5	+1.8 / −1.8	+6.6 / −6.2	+2.6 / −2.8	+2.5 / −2.5
	97–133	1.1×10^4	+7.7 / −8.8	+0.2 / −0.2	+0.0 / −0.0	+0.5 / −0.2	+3.8 / −5.0	+1.1 / +1.1	+0.8 / +0.2	+2.1 / −1.9	+4.7 / −4.4	+3.3 / −3.7	+2.5 / −2.5
	133–174	3.3×10^3	+9.6 / −6	+0.2 / −0.2	+0.0 / −0.0	+0.0 / −0.0	+0.4 / −2.8	+8.4 / −2.1	+0.2 / −0.2	+1.8 / −2.0	+1.8 / −2.1	+2.7 / −3.0	+2.5 / −2.5
	174–220	1.1×10^3	+5.5 / −9	+0.0 / −0.0	+0.0 / −0.0	+0.4 / −0.3	+0.3 / −0.7	+1.8 / −6.7	+0.1 / −0.5	+3.0 / −3.4	+1.9 / −2.5	+2.7 / −3.1	+2.5 / −2.5
	220–272	3.9×10^2	+5.8 / −6.8	+0.0 / −0.0	+0.0 / −0.0	+0.3 / −0.9	+0.2 / −0.4	+3.1 / −2.8	+0.5 / −1.0	+2.2 / −3.2	+1.2 / −2.3	+3.3 / −3.8	+2.5 / −2.5
	272–395	1.8×10^2	+12 / −14	+0.0 / −0.0	+0.0 / −0.0	+1.0 / −0.9	+0.4 / −0.0	+3.0 / −3.3	+0.2 / −1.0	+5.3 / −6.1	+7.6 / −8.7	−3.8 / −7.0	+2.5 / −2.5
	395–548	32	+11 / −11	+0.0 / −0.0	+0.0 / −0.0	+1.0 / −2.1	+0.7 / −0.0	+3.2 / −3.1	+0.0 / −1.0	+5.0 / −6.1	+7.3 / −8.7	+5.1 / −5.9	+2.5 / −2.5
	548–737	5.4	+21 / −21	+0.0 / −0.0	+0.0 / −0.0	+2.1 / −5.8	+0.0 / −0.1	+1.9 / −0.6	+0.4 / +0.2	+10.1 / −9.1	+15.2 / −15.3	+10.0 / −11.7	+2.5 / −2.5
	737–1101	1.2	+43 / −44	+0.0 / −0.0	+0.0 / −0.0	+5.8 / −6.0	+0.1 / −0.0	+2.0 / −3.0	+0.2 / −0.3	+17.2 / −15.3	+35.3 / −36.0	+16.3 / −19.1	+2.5 / −2.5
	1101–1588	0.1	$+1.1 \times 10^2$ / $−1.2 \times 10^2$	+0.2 / −0.2	+0.0 / −0.0	+8.1 / −8.6	+0.0 / −0.3	+3.5 / −2.5	+0.5 / −0.0	+38.1 / −30.5	+103.0 / −110.1	+18.3 / −21.7	+2.5 / −2.5
$1.0 < |y| < 1.5$	74–97	2×10^4	+19 / −15	+9.2 / −9.3	+0.0 / −0.0	+0.1 / −0.3	+10.3 / −0.9	+10.2 / −7.6	+0.6 / −0.0	+1.6 / −1.7	+7.2 / −7.1	+2.4 / −2.6	+2.5 / −2.5
	97–133	8.9×10^3	+11 / −13	+1.8 / −1.7	+0.0 / −0.0	+0.8 / −0.4	+7.6 / −8.8	+1.1 / −6.3	+1.5 / −0.1	+2.3 / −1.8	+5.3 / −4.5	+3.1 / −3.4	+2.5 / −2.5
	133–174	2.5×10^3	+12 / −10	+1.4 / −1.4	+0.0 / −0.0	+0.8 / −2.0	+3.0 / −0.6	+9.5 / −6.6	+0.5 / −1.3	+1.3 / −2.4	+3.6 / −4.8	+3.3 / −3.6	+2.5 / −2.5

(continued)

Table 9.3 (continued)

y	p_T (GeV)	Cross section	Tot. syst. unc. (%)	Back-ground (%)	Regulari-sation (%)	PU re-weighting (%)	Over-weighted PU (%)	JEC (%)	JER (%)	Heavy-flavour SF (%)	Light-flavour SF (%)	Model reweighting (%)	Lumi (%)
	174–220	7.9×10^2	+10 / −8	+0.2 / −0.1	+0.0 / −0.0	+2.4 / −0.5	+0.6 / −0.0	+2.9 / −4.7	+1.2 / −1.1	+4.9 / −1.9	+6.4 / −3.5	+3.8 / −4.3	+2.5 / −2.5
	220–272	2.6×10^2	+12 / −8.8	+0.1 / −0.0	+0.0 / −0.0	+2.5 / −1.6	+0.3 / −0.4	+6.2 / −5.4	+2.7 / −1.3	+4.7 / −0.5	+7.2 / −3.0	+4.6 / −5.3	+2.5 / −2.5
	272–395	1.2×10^2	+26 / −17	+0.0 / −0.0	+0.0 / −0.0	+6.9 / −0.1	+0.2 / −0.0	+2.9 / −2.6	−3.6 / −3.4	+9.8 / −2.4	+20.9 / −14.0	+7.3 / −8.4	+2.5 / −2.5
	395–548	20	+27 / −20	+0.0 / −0.0	+0.0 / −0.0	+9.6 / −4.3	+0.1 / −0.1	+3.8 / −3.0	+2.8 / +0.8	+8.4 / −2.3	+22.3 / −16.8	+7.7 / −8.9	+2.5 / −2.5
	548–737	3.4	+41 / −46	+0.0 / −0.0	+0.0 / −0.0	+4.5 / −10.4	+0.1 / −0.3	+0.8 / −2.3	+0.8 / −1.5	+9.0 / −10.6	+36.9 / −40.3	+13.0 / −14.7	+2.5 / −2.5
	737–1101	0.62	+90 / −92	+0.0 / −0.0	+0.0 / −0.0	+17.7 / −17.8	+1.1 / −0.3	+1.1 / −0.9	+1.0 / −0.4	+18.9 / −15.4	+83.3 / −85.8	+21.3 / −24.6	+2.5 / −2.5
	1101–1588	0.063	$+1.4\times10^2$ / $−1.5\times10^2$	+0.4 / −0.4	+0.0 / −0.0	+12.4 / −12.4	+0.7 / −0.3	+0.1 / −1.2	+1.5 / −1.4	+40.3 / −31.4	+130.9 / −148.6	+23.3 / −27.6	+2.5 / −2.5
1.5 < |y| < 2.0	74–97	1.8×10^4	+16 / −15	+7.0 / −7.0	+0.0 / −0.0	+0.1 / −0.5	+6.4 / −3.8	+10.6 / −10.9	+0.4 / −0.9	+1.8 / −2.0	+5.7 / −5.8	+2.3 / −2.6	+2.5 / −2.5
	97–133	6.8×10^3	+8.9 / −9.5	+1.3 / −1.3	+0.0 / −0.0	+0.9 / −0.3	+2.5 / −2.8	+5.0 / −5.0	+0.7 / −0.0	+2.5 / −2.0	+4.9 / −4.5	+3.0 / −3.4	+2.5 / −2.5
	133–174	1.9×10^3	+15 / −15	+0.7 / −0.7	+0.0 / −0.0	+1.2 / −0.9	+0.3 / −0.0	+11.3 / −5.7	+0.7 / −0.3	+2.3 / −2.0	+6.9 / −6.6	+5.0 / −5.6	+2.5 / −2.5
	174–220	4.9×10^2	+9.9 / −9.4	+0.2 / −0.1	+0.0 / −0.0	+2.0 / −2.9	+1.2 / −0.5	+5.6 / −11.9	+7.5 / −8.7	+3.0 / −3.6	+5.3 / −6.1	+4.2 / −4.8	+2.5 / −2.5
	220–272	1.8×10^2	+15 / −26	+0.1 / −0.1	+0.0 / −0.0	+7.6 / −9.4	+0.1 / −0.0	+7.2 / −8.9	+4.8 / −4.4	+5.7 / −11.4	+23.1 / −15.7	+5.0 / −5.7	+2.5 / −2.5
	272–395	75	+31 / −18	+0.0 / −0.0	+0.0 / −0.0	+11.6 / −1.4	−0.2 / −0.3	+5.6 / −5.9	+0.2 / −0.6	+12.3 / −1.3	+23.1 / −12.6	+7.8 / −9.2	+2.5 / −2.5
	395–548	9	+30 / −35	+0.0 / −0.0	+0.0 / −0.0	+7.4 / −11.1	+0.1 / −0.1	+6.8 / −5.9	+2.3 / −0.5	+5.2 / −7.6	+25.7 / −28.7	+10.5 / −12.3	+2.5 / −2.5
	548–737	1.1	+56 / −56	+0.0 / −0.0	+0.0 / −0.0	+18.3 / −18.3	−0.1 / −0.1	+9.9 / −11.9	+2.3 / −2.7	+11.6 / −9.2	+49.5 / −48.9	+14.0 / −16.1	+2.5 / −2.5
	737–1101	0.14	+79 / −82	+0.5 / −0.5	+0.0 / −0.0	+23.5 / −25.6	−0.1 / −0.1	+9.9 / −11.9	+2.7 / −2.7	+16.0 / −14.7	+72.1 / −73.8	+13.5 / −16.9	+2.5 / −2.5
2.0 < |y| < 2.4	74–97	1.1×10^4	+17 / −14	+5.7 / −5.5	+0.0 / −0.0	+1.2 / −1.5	+9.4 / −6.8	+7.8 / −8.1	+2.7 / −1.5	+2.0 / −2.2	+7.7 / −7.6	+3.8 / −4.3	+2.5 / −2.5
	97–133	4.5×10^3	+12 / −8	+0.8 / −0.9	+0.0 / −0.0	+0.5 / −0.3	+1.1 / −0.3	+9.3 / −4.3	+1.8 / −1.4	+2.5 / −1.6	+5.3 / −4.6	+3.1 / −3.5	+2.5 / −2.5
	133–174	1×10^3	+16 / −20	+0.8 / −0.7	+0.0 / −0.0	+1.6 / −4.6	−0.6 / −0.3	+8.9 / −9.5	+0.8 / −2.2	+0.9 / −3.5	+10.4 / −12.9	+6.5 / −7.1	+2.5 / −2.5
	174–220	2.7×10^2	+27 / −14	+0.2 / −0.2	+0.0 / −0.0	+12.0 / −3.8	+0.9 / −0.3	+3.6 / −8.2	+9.0 / −5.9	+11.1 / −5.1	+18.9 / −3.0	+5.0 / −5.6	+2.5 / −2.5
	220–272	63	+27 / −41	+0.1 / −0.1	+0.0 / −0.0	+1.5 / −13.8	−0.3 / −0.2	+16.4 / −6.7	+6.4 / −7.2	+3.5 / −11.2	+19.4 / −34.4	+7.0 / −8.0	+2.5 / −2.5
	272–395	26	+37 / −46	+0.1 / −0.1	+0.0 / −0.0	+10.8 / −17.1	−0.3 / −0.2	+0.5 / −7.3	+0.9 / −4.5	+4.5 / −9.2	+33.2 / −38.7	+10.9 / −12.9	+2.5 / −2.5
	395–548	2.7	+53 / −39	+0.0 / −0.0	+0.0 / −0.0	+20.6 / −12.5	+0.3 / −0.2	+19.8 / −10.1	+11.4 / −8.1	+12.0 / −2.2	+40.0 / −31.0	+11.6 / −14.1	+2.5 / −2.5
	548–737	0.25	+38 / −45	+0.2 / −0.2	+0.0 / −0.0	+8.3 / −14.6	−0.2 / −0.2	+7.8 / −3.1	+2.5 / −10.1	+7.7 / −11.0	+33.4 / −37.7	+10.8 / −13.2	+2.5 / −2.5

Table 9.4 Fraction of b jets in the inclusive jet measurement and systematic uncertainties

y	p_T (GeV)	Fraction	Tot. syst. unc. (%)	Background (%)	Regularisation (%)	PU re-weighting (%)	Over-weighted PU (%)	JEC (%)	JER (%)	Heavy-flavour SF (%)	Light-flavour SF (%)	Model reweighting (%)
$\|y\| < 0.5$	74–97	0.036	+9.8/−8	+1.8/−2.0	+0.0/−0.0	+0.5/−0.4	+6.7/−1.5	+0.3/−3.3	+0.3/−0.1	+1.7/−1.6	+5.7/−5.5	+3.5/−3.7
	97–133	0.041	+5.6/−6.8	+0.2/−0.1	+0.0/−0.0	+0.1/−0.2	+2.4/−4.6	+1.6/−1.0	−0.1/−0.2	+2.0/−2.1	+3.3/−3.3	+2.7/−2.9
	133–174	0.046	+4.2/−7.5	+0.1/−0.1	+0.0/−0.0	+0.4/−0.2	+0.9/−5.9	+1.0/−2.4	+0.4/−0.2	+1.9/−1.8	+2.3/−2.5	+2.5/−2.8
	174–220	0.045	+5.7/−4.6	+0.0/−0.0	+0.0/−0.0	+0.8/−0.1	+1.4/+0.4	+1.7/−1.0	+0.4/−0.1	+3.6/−2.7	+2.5/−1.7	+2.8/−3.2
	220–272	0.047	+4.5/−7.7	+0.0/−0.0	+0.0/−0.0	+1.2/−1.5	+0.4/−0.3	+0.6/−2.2	+1.0/−1.4	+1.3/−3.8	+1.8/−4.5	+3.5/−3.9
	272–395	0.048	+9.1/−9.2	+0.0/−0.0	+0.0/−0.0	+0.8/−0.6	+0.2/−0.1	+1.2/−0.2	+0.1/+0.1	+5.5/−5.0	+4.8/−4.6	+5.3/−6.0
	395–548	0.048	+10/−9.6	+0.0/−0.0	+0.0/−0.0	+0.9/−0.2	+0.1/−0.1	+0.1/+0.4	+0.5/−0.2	+5.2/−3.8	+7.3/−6.2	+5.3/−6.2
	548–737	0.048	+16/−16	+0.0/−0.0	+0.0/−0.0	+0.7/−0.7	+0.1/−0.0	+0.4/−1.3	+0.0/−0.5	+8.8/−8.0	+11.0/−11.0	+7.8/−9.2
	737–1101	0.049	+29/−30	+0.0/−0.0	+0.0/−0.0	+1.6/−1.6	+0.1/−0.1	+1.1/−0.6	+0.1/−0.0	+14.4/−12.8	+22.4/−22.9	+12.1/−14.2
	1101–1588	0.049	+82/−86	+0.1/−0.1	+0.0/−0.0	+3.1/−3.1	+0.3/−0.2	+0.7/−2.2	+0.0/−0.2	+32.0/−25.5	+74.2/−79.8	+15.4/−17.9
$0.5 < \|y\| < 1.0$	74–97	0.034	+9.6/−7.4	+1.6/−1.9	+0.0/−0.0	+0.4/−0.6	+6.1/−0.0	+0.2/−1.3	+0.2/−0.2	+1.8/−1.8	+6.5/−6.2	+2.6/−2.8
	97–133	0.039	+6.5/−7.4	+0.3/−0.3	+0.0/−0.0	+0.4/−0.1	+0.0/−1.7	+1.1/−1.1	+0.6/−0.1	+2.1/−1.9	+4.7/−4.4	+3.3/−3.6
	133–174	0.042	+5.5/−4.8	+0.2/−0.2	+0.0/−0.0	+0.1/−0.3	+0.1/−2.9	+4.0/−3.2	+0.0/−0.1	+1.8/−2.1	+1.8/−4.4	+2.7/−3.0
	174–220	0.045	+5.6/−7.2	+0.0/−0.0	+0.0/−0.0	+0.4/−0.7	+0.0/−1.3	−1.8/+3.2	−0.5/+0.1	+3.0/−3.3	+2.0/−2.4	+3.3/−3.1
	220–272	0.047	+4.2/−5.8	+0.1/−0.1	+0.0/−0.0	+0.4/−0.4	+0.4/−0.3	−5.0/−0.7	+0.7/−0.3	+2.1/−3.3	+1.1/−2.4	+3.3/−3.8
	272–395	0.042	+11/−13	+0.0/−0.0	+0.0/−0.0	+1.1/−2.0	+0.5/+0.3	+0.7/+0.1	+0.7/−0.9	+5.4/−6.0	+7.6/−8.6	+6.0/−7.0
	395–548	0.045	+10/−10	+0.0/−0.0	+0.0/−0.0	+1.1/−1.1	+0.3/−0.6	+0.1/−0.3	+0.3/−0.8	+5.0/−4.5	+7.2/−7.0	+5.1/−5.9
	548–737	0.04	+21/−21	+0.0/−0.0	+0.0/−0.0	+2.2/−2.2	+0.0/−0.0	+2.0/−0.4	+0.1/−0.1	+10.1/−9.1	+15.1/−15.1	+10.0/−11.6
	737–1101	0.038	+43/−44	+0.0/−0.0	+0.0/−0.0	+5.8/−6.0	+0.3/+0.1	+0.1/−0.1	+0.1/−0.4	+17.2/−15.3	+35.0/−35.9	+16.3/−19.0
	1101–1588	0.039	+1.1×10²/−1.2×10²	+0.1/−0.1	+0.0/−0.0	+8.1/−8.6	+0.0/−0.4	+2.1/−1.1	+0.7/−0.2	+38.0/−30.4	+101.8/−110.1	+18.2/−21.6
$1.0 < \|y\| < 1.5$	74–97	0.032	+11/−9.2	+3.7/−4.4	+0.0/−0.0	+0.1/−0.4	+6.7/−4.9	+2.4/−0.6	+0.0/−2.2	+1.6/−1.7	+7.2/−7.1	+2.4/−2.6
	97–133	0.035	+8.2/−11	+0.8/−0.7	+0.0/−0.0	+0.9/−0.5	+4.9/−6.3	+0.4/−7.1	+0.9/−0.6	+2.3/−1.9	+5.2/−4.5	+3.1/−3.4
	133–174	0.038	+7.8/−6.8	+0.7/−0.7	+0.0/−0.0	+0.9/−1.9	+4.5/−0.3	+3.7/−1.3	+0.6/−0.8	+1.4/−2.3	+3.6/−4.6	+3.3/−3.6

(continued)

Table 9.4 (continued)

y	p_T(GeV)	Fraction	Tot. syst. unc. (%)	Back-ground (%)	Regulari-sation (%)	PU re-weighting (%)	Over-weighted PU (%)	JEC (%)	JER (%)	Heavy-flavour SF (%)	Light-flavour SF (%)	Model reweighting (%)
	174–220	0.039	+8.9 / −6.5	+0.1 / −0.1	+0.0 / −0.0	+2.1 / −0.4	+0.4 / −0.0	+0.2 / −2.4	+1.2 / −0.7	+4.7 / −2.0	+6.1 / −3.6	+3.8 / −4.2
	220–272	0.037	+11 / −6.5	+0.0 / −0.0	+0.0 / −0.0	+2.6 / −1.8	+0.5 / −0.6	+1.2 / −0.7	+2.5 / −1.7	+4.9 / −0.4	+7.3 / −2.8	+4.6 / −5.3
	272–395	0.034	+25 / −17	+0.0 / −0.0	+0.0 / −0.0	+6.8 / −0.1	+0.3 / −0.2	+1.5 / −1.5	+3.8 / −3.1	+9.8 / −2.4	+20.7 / −13.9	+7.3 / −8.4
	395–548	0.037	+27 / −20	+0.0 / −0.0	+0.0 / −0.0	+9.5 / −4.4	+0.0 / −0.2	+0.5 / −0.1	+3.0 / −2.5	+8.3 / −2.4	+22.1 / −16.8	+7.8 / −8.9
	548–737	0.036	+41 / −45	+0.0 / −0.0	+0.0 / −0.0	+7.5 / −10.4	+0.1 / −0.4	+0.7 / −2.3	+0.6 / −1.7	+9.0 / −10.6	+36.5 / −40.1	+13.0 / −14.7
	737–1101	0.033	+89 / −92	+0.0 / −0.0	+0.0 / −0.0	+17.6 / −17.8	+0.0 / −0.3	+4.8 / −2.6	+1.2 / −0.7	+18.8 / −15.4	+82.5 / −85.8	+21.2 / −24.6
	1101–1588	0.056	+1.4×10² / −1.6×10²	+0.2 / −0.2	+0.0 / −0.0	+12.4 / −12.5	+0.9 / −0.3	+5.5 / −6.7	+1.9 / −1.9	+40.2 / −31.4	+128.8 / −149.1	+23.2 / −27.6
$1.5 < \|y\| < 2.0$	74–97	0.032	+7.3 / −7.2	+2.0 / −2.3	+0.0 / −0.0	+0.1 / −0.2	+2.8 / −0.5	+0.5 / −1.5	+0.2 / −0.6	+1.8 / −1.9	+5.7 / −5.7	+2.3 / −2.6
	97–133	0.031	+6.8 / −6.9	+0.5 / −0.5	+0.0 / −0.0	+0.7 / −0.2	+0.4 / −1.3	+2.5 / −3.2	+0.5 / −0.1	+2.4 / −2.3	+4.9 / −4.5	+3.0 / −3.4
	133–174	0.032	+9.3 / −10	+0.2 / −0.2	+0.0 / −0.0	+1.2 / −0.8	+1.0 / −1.3	+2.7 / −4.8	+0.5 / −0.2	+2.3 / −1.9	+6.9 / −6.5	+4.9 / −5.6
	174–220	0.029	+11 / −9.3	+0.2 / −0.2	+0.0 / −0.0	+1.8 / −2.8	+1.7 / −0.2	+7.2 / −2.1	+0.8 / −0.8	+3.0 / −3.6	+5.3 / −6.1	+4.2 / −4.8
	220–272	0.033	+13 / −24	+0.1 / −0.1	+0.0 / −0.0	+7.8 / −9.5	+0.0 / −0.3	+0.1 / −2.3	+7.2 / −9.2	+5.8 / −11.5	+1.6 / −15.8	+5.0 / −5.7
	272–395	0.028	+30 / −16	+0.0 / −0.0	+0.0 / −0.0	+11.5 / −1.6	+0.0 / −0.3	+0.4 / −1.2	+5.0 / −3.9	+12.2 / −1.4	+22.8 / −12.6	+7.8 / −9.1
	395–548	0.024	+29 / −34	+0.0 / −0.0	+0.0 / −0.0	+7.5 / −11.0	+0.2 / −0.1	+0.7 / −0.7	+0.2 / −0.8	+5.3 / −7.5	+25.7 / −28.6	+10.5 / −12.3
	548–737	0.021	+56 / −55	+0.0 / −0.0	+0.0 / −0.0	+18.5 / −18.1	+0.1 / −0.0	+0.7 / −0.2	+2.8 / −0.5	+11.8 / −9.0	+49.6 / −48.7	+14.0 / −16.1
	737–1101	0.019	+78 / −81	+0.4 / −0.4	+0.0 / −0.0	+23.5 / −25.7	+0.0 / −0.2	+0.2 / −3.5	+2.8 / −3.1	+15.9 / −14.7	+71.5 / −73.8	+13.5 / −16.9
$2.0 < \|y\| < 2.4$	74–97	0.029	+11 / −9.5	+0.8 / −0.9	+0.0 / −0.0	+0.5 / −0.7	+5.5 / −0.1	+1.6 / −2.7	+0.9 / −0.3	+2.0 / −2.2	+7.7 / −7.6	+3.8 / −4.3
	97–133	0.03	+8.3 / −6.6	+0.3 / −0.4	+0.0 / −0.0	+0.8 / −0.1	+3.4 / −2.3	+3.6 / −1.2	+0.5 / −0.2	+2.4 / −1.7	+5.3 / −4.6	+3.1 / −3.5
	133–174	0.028	+12 / −17	+0.2 / −0.1	+0.0 / −0.0	+1.4 / −4.3	+0.4 / −5.8	+1.0 / −2.7	+1.2 / −1.6	+1.0 / −3.5	+10.4 / −12.8	+6.5 / −7.1
	174–220	0.027	+26 / −12	+0.3 / −0.3	+0.0 / −0.0	+11.6 / −3.8	+2.6 / −0.7	+1.9 / −3.2	+7.9 / −6.7	+10.9 / −4.9	+18.7 / −3.1	+5.0 / −5.6
	220–272	0.02	+24 / −40	+0.1 / −0.1	+0.0 / −0.0	+1.3 / −13.7	+0.0 / −0.4	+8.8 / −0.1	+6.2 / −6.9	+4.3 / −11.0	+19.5 / −34.2	+7.0 / −8.0
	272–395	0.019	+37 / −46	+0.2 / −0.2	+0.0 / −0.0	+10.7 / −17.1	+0.4 / −0.3	+1.1 / −5.7	+0.2 / −5.3	+4.4 / −9.2	+33.0 / −38.7	+10.8 / −12.9
	395–548	0.019	+51 / −38	+0.0 / −0.0	+0.0 / −0.0	+20.7 / −12.3	+0.0 / −0.3	+12.0 / −3.8	+10.5 / −9.2	+12.1 / −2.1	+40.0 / −30.9	+11.6 / −14.2
	548–737	0.02	+37 / −45	+0.1 / −0.0	+0.0 / −0.0	+8.5 / −14.7	+0.0 / −0.3	+5.4 / −1.2	+3.5 / −11.1	+7.7 / −11.0	+33.2 / −37.5	+10.8 / −13.2

Table 9.5 Parameters of the tune for UE in PYTHIA 8

Tune	Parameter	Value
ee tune 7	StringFlav:probStoUD	0.217
	StringFlav:probQQtoQ	0.081
	StringFlav:probSQtoQQ	0.915
	StringFlav:probQQ1toQQ0	0.0275
	StringFlav:mesonUDvector	0.50
	StringFlav:mesonSvector	0.55
	StringFlav:mesonCvector	0.88
	StringFlav:mesonBvector	2.20
	StringFlav:etaSup	0.60
	StringFlav:etaPrimeSup	0.12
	StringFlav:popcornSpair	0.90
	StringFlav:popcornSmeson	0.50
	StringFlav:suppressLeadingB	false
	StringZ:aLund	0.68
	StringZ:bLund	0.98
	StringZ:aExtraSquark	0.00
	StringZ:aExtraDiquark	0.97
	StringZ:rFactC	1.32
	StringPT:sigma	0.335
	StringPT:enhancedFraction	0.01
	StringPT:enhancedWidth	2.0
	TimeShower:alphaSvalue	0.1365
	TimeShower:alphaSorder	1
	TimeShower:alphaSuseCMW	False
	TimeShower:pTmin	0.5
	TimeShower:pTminChgQ	0.5
pp tune 14	PDF:pSet	NNPDF
	SigmaProcess:alphaSvalue	0.130
	SigmaTotal:zeroAXB	True
	SigmaDiffractive:dampen	True
	SigmaDiffractive:maxXB	65.0
	SigmaDiffractive:maxAX	65.0
	SigmaDiffractive:maxXX	65.0
	Diffraction:largeMassSuppress	4.0
	TimeShower:dampenBeamRecoil	True
	TimeShower:phiPolAsym	True
	SpaceShower:alphaSvalue	0.1365
	SpaceShower:alphaSorder	1
	SpaceShower:alphaSuseCMW	False
	SpaceShower:samePTasMPI	False

(continued)

Table 9.5 (continued)

Tune	Parameter	Value
	`SpaceShower:pT0Ref`	2.0
	`SpaceShower:ecmRef`	7000.0
	`SpaceShower:ecmPow`	0.0
	`SpaceShower:pTmaxFudge`	1.0
	`SpaceShower:pTdampFudge`	1.0
	`SpaceShower:rapidityOrder`	True
	`SpaceShower:rapidityOrderMPI`	True
	`SpaceShower:phiPolAsym`	True
	`SpaceShower:phiIntAsym`	True
	`MultipartonInteractions:alphaSvalue`	0.130
	`MultipartonInteractions:bProfile`	3
	`MultipartonInteractions:expPow`	1.85
	`MultipartonInteractions:a1`	0.15
	`BeamRemnants:primordialKTsoft`	0.9
	`BeamRemnants:primordialKThard`	1.8
	`BeamRemnants:halfScaleForKT`	1.5
	`BeamRemnants:halfMassForKT`	1.0
	`ColourReconnection:mode`	0
	`ColourReconnection:range`	1.80
CUETP8M1	`MultipartonInteractions:pT0Ref`	2.4024
	`MultipartonInteractions:ecmPow`	0.25208
	`MultipartonInteractions:expPow`	1.6
	`StringZ:rFactB`	0.895

References

1. Khachatryan V et al (2016) Measurement of the double-differential inclusive jet cross section in proton – proton collisions at $\sqrt{s} = 13$ TeV. Eur Phys J C76(8):451. https://doi.org/10.1140/epjc/s10052-016-4286-3 arXiv:1605.04436 [hep-ex]
2. Khachatryan V et al (2017) Measurement and QCD analysis of double-differential inclusive jet cross sections in pp collisions at $\sqrt{s} = 8$ TeV and cross section ratios to 2.76 and 7 TeV. JHEP 03:156. https://doi.org/10.1007/JHEP03(2017)156, arXiv:1609.05331 [hep-ex]
3. Chatrchyan S et al (2012) Inclusive b-jet production in pp collisions at $\sqrt{s} = 7$ TeV. JHEP 04:084. https://doi.org/10.1007/JHEP04(2012)084, arXiv:1202.4617 [hep-ex]
4. Aad G et al (2011) Measurement of the inclusive and dijet cross-sections of b^- jets in pp collisions at $\sqrt{s} = 7$ TeV with the ATLAS detector. Eur Phys J C71:1846. https://doi.org/10.1140/epjc/s10052-011-1846-4, arXiv:1109.6833 [hep-ph]
5. Alioli S et al (2011) Jet pair production in POWHEG. JHEP 1104:081. https://doi.org/10.1007/JHEP04(2011)081, arXiv:1012.3380
6. Alioli S et al (2010) A general framework for implementing NLO calculations in shower Monte Carlo programs: the POWHEG BOX. JHEP 06:043. https://doi.org/10.1007/JHEP06(2010)043, arXiv:1002.2581 [hep-ph]

7. Nason PA (2010) Recent developments in POWHEG. In: Proceedings, 9th international symposium on radiative corrections: applications of quantum field theory to phenomenology. (RAD-COR 2009). vol RADCOR2009, p 018. arXiv:1001.2747 [hep-ph]
8. Skands P, Carrazza S, Rojo J (2013) Tuning PYTHIA 8.1: the Monash 2013 Tune. Eur Phys J C74(8):3024. https://doi.org/10.1140/epjc/s10052-014-3024-y, arXiv:1404.5630 [hep-ph]

Part III
Conclusions

Chapter 10
Summary, Conclusions and Perspectives

The measurement that has been presented in this thesis is a textbook measurement. It is one of the most important observables in proton-proton collisions.

In this chapter, we summarise the analysis. In addition, we discuss additional results that can be obtained thanks to the techniques explained in this thesis. In this section, limitations and successes of these techniques are discussed and prospects for future measurements are given.

10.1 Summary

In this thesis, we have presented a new measurement of the inclusive b jet production at $\sqrt{s} = 13$ TeV.

After a long introduction to situate the context of the measurement, we showed how to obtain the signal with the CMS experiment, and we corrected it to parton level. For this, we applied an advanced method of unfolding to disentangle the b jets from the light and c jets. Then we compared it to predictions at LO and at NLO; in addition, we also measured and compared the fraction of b jets in the inclusive jet production.

We also investigated the mechanisms of productions of b jets. The contribution from PS dominate at high p_T.

The cross section is measured over six orders of magnitude, with a large coverage of the phase space. The transverse momentum goes from 100 GeV to 1 TeV and the rapidity from -2.4 to 2.4. The agreement with the theory is a great success.

Its validation at LHC opens up many new measurements. For instance, the improvement of the modelling of $b\bar{b}$ background could lead to the measurement the Higgs boson in the $H \rightarrow b\bar{b}$ channel.

© Springer Nature Switzerland AG 2019
P. Connor, *Inclusive b Jet Production in Proton-Proton Collisions*, Springer Theses,
https://doi.org/10.1007/978-3-030-34383-5_10

10.2   Perspectives

Many issues still need to be solved in Quantum Chromodynamics (QCD). As we saw, despite the numerous successes of the SM, difficulties remain in the description of the MPI, or in general, of most non-perturbative aspects.

We review here some perspectives of new measurements and new techniques of measurement with b jets and in which circumstances they may be of interest.

10.2.1   New Measurements with Run-II Data

We discuss here HF measurements that can be done with $\sqrt{s} = 13$ TeV data at LHC.

10.2.1.1   Cross Section for $b\bar{b}$ Production

The measurement of $b\bar{b}$ is the natural next step, after the measurement of inclusive b jet production.

First, we can measure the mass spectrum of the pair, and investigate evolution effects by reconstructing extra radiations.

Then, we can measure the triple differential cross section of the $b\bar{b}$ dijet system as a function of the average transverse momentum, of the rapidity boost and of the rapidity separation. The interest of this measurement would be to constrain PDFs; it is also the textbook measurement to describe the ME of production of $b\bar{b}$..

In addition, we can study the azimuthal correlations among b jets, and also with additional jet production. While the inclusive b jet measurement is more sensitive to the ME, the azimuthal correlations with additional radiations having significant p_T is more sensitive to the UE. One could study effect more related to the PS and MPI.

10.2.1.2   Inclusive c Jet Cross Section

The inclusive c jet cross section would be another good test of the flavour democracy of QCD. A similar measurement has never been done at LHC and would be a premiere.

However, although it looks similar to the inclusive b jet measurement, as was done in this thesis, it is slightly more complicated. Indeed, a c tagger consists of a double tagger: an "anti-light" tagger plus an "anti-b" tagger. At the time of writing the thesis, recent developments have been achieved regarding c tagging [1]. While with the first versions of the tools, the analysis could not be reproduced with c jets, the performance is significantly improving.

The strategy to measure a c jet cross section would be to disentangle simultaneously light, c and b jets For this, we can use the double tagger or DeepCSV, and apply similar techniques as what has been described in this thesis, extending the matrix in Eq. 8.2 to a 3×3 matrix.

10.2.2 Techniques of Measurements

The use of Deep Learning techniques is getting more and more widespread in the context of b tagging. The successor of `DeepCSV`, including more observables and more sophisticated, is already under development and will be available for 2017 and 2018 data, called `DeepFlavour`.

In addition, a novel technique of tagging for silicon detectors has recently been published and would be particularly adapted to CMS [2]. It is illustrated in Fig. 10.1. Currently, as we described in this volume, the tagging techniques are mostly based on tracking and vertexing. In this new method, instead of losing efficiency at tracking and vertexing to perform b tagging, patterns of hits in the tracker are directly investigated. This technique only starts to be efficient for boosted B's, and should be included in the techniques of Deep Learning used in current taggers. This could significantly improve the purity at high p_T.

Moreover, in the coming decade, the CMS tracker should extended down to $y = 4.0$; with, in addition, the ever increasing pile-up conditions, this technique could significantly help to perform b tagging in an environment where tracks are not evident to reconstruct.

Fig. 10.1 Sketch of the decay of a B hadron between to layers of the tracker. The gray planes represents silicon modules; the arrows represent tracks; the hits represent hits. Figure reproduced with permission of author [2]

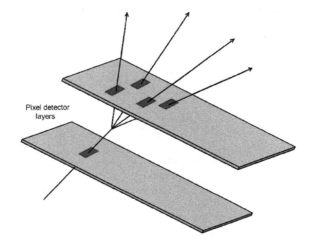

Pixel detector layers

10.2.3   Prospects at Longer Term

At longer term, the LHC will undergo substantial upgrades in order to deliver higher luminosities [3], aiming at an integrated luminosity of 3000 fb^{-1}. Discussions are also ongoing, whether LHC could be upgraded for higher energies in the centre-of-mass system, with $\sqrt{s} \approx 33$ TeV. This implies that the statistics at the TeV scale will become significantly larger, and that multi-differential precision measurements will be possible in this region of the phase space.

In the context of b jets, this has several implications: first, it is important to pursue effort in b tagging techniques. Second, as extra radiations are extremely important at high p_T, the modelling of PS has to be refined further, including additional orders.

But the most exciting comes in terms of physics. It will also become possible to test the flavour democracy by comparing bottom and top productions, since the scale will be much larger than the mass of the top quark. Indeed, any deviations from the flavour democracy might be a hint to new physics.

References

1. Sirunyan AM et al (2017) Identification of heavy-flavour jets with the CMS detector in pp collisions at 13 TeV. arXiv:1712.07158 [physics.ins-det]
2. Huffman BT, Jackson C, Tseng J (2016) Tagging b quarks at extreme energies without tracks. J Phys G43(8):085001. https://doi.org/10.1088/0954-3899/43/8/085001, arXiv:1604.05036 [hep-ex]
3. Apollinari G et al (2017) High-luminosity large hadron collider (HL-LHC): technical design report V. 0.1. CERN yellow reports: monographs. Geneva: CERN. https://cds.cern.ch/record/2284929

Appendix
Tracker Alignment

AS DESCRIBED PREVIOUSLY in Chap. 3, the central CMS tracker consists of several superimposed layers of silicon modules (as of 2016). In order to achieve reconstruction of tracks with optimal precision despite the finite fabrication tolerances of the large structures, the modules of the tracking system need to be *aligned*. Corrections for the position, orientation and curvature need to be computed for every single sensor, possibly changing with time.

The strategy to align the modules is based on the reconstruction of tracks. One technique consists in determining the parameters of the modules and of the tracks simultaneously, involving up to millions of parameters to fit simultaneously.

In the context of this thesis, it is important to mention the alignment is crucial for the good performance of b tagging, since it relies on the reconstruction of tracks and SVs.

In this chapter, we explain how to solve the challenging problem of alignment with the track-based approach. We first describe the general strategy adopted at CMS (Sect. A.1); then we present the *Legacy Alignment* of the 2016 data (Sect. A.2).

A significant part of the work spent at DESY for this thesis and for the CMS collaboration was devoted to the alignment of the tracker. This appendix is the written version of a talk given at the TIPP 2017 conference in Beijing.

A.1 Introduction

The purpose of a silicon tracker is to reconstruct the tracks from their hits in the successively traversed modules; the basic principles are illustrated in Fig. A.1a. At CMS, the algorithm for the track reconstruction is the *Kálmán filter*, as described in Sect. 3.2.2.1.

At the mounting of the tracker, the precision of the mechanical alignment is typically of $\mathcal{O}(0.1\,\text{mm})$. At this stage, the uncertainty on the alignment is much larger than the hit resolution:

© Springer Nature Switzerland AG 2019
P. L. S. Connor, *Inclusive b Jet Production in Proton-Proton Collisions*, Springer Theses,
https://doi.org/10.1007/978-3-030-34383-5

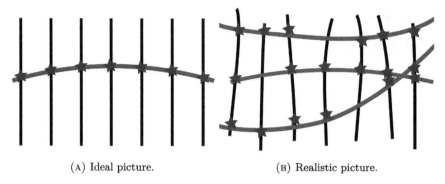

(A) Ideal picture.   (B) Realistic picture.

Fig. A.1 Track reconstruction in the context of alignment. The black, straight lines represent the **silicon modules**, seen transversally; the dark blue, curved line represents tracks; the red stars represent the track hits

$$\sigma_{\text{align}} \gg \sigma_{\text{hit}} \tag{A.1}$$

Therefore, it is important to improve the alignment to a resolution of (at most) the same order as the hit resolution:

$$\sigma_{\text{align}} \lesssim \sigma_{\text{hit}} \tag{A.2}$$

Moreover, while a random misalignment only degrades the accuracy of the measurement, a systematic misalignment may lead to a bias in physics results. Tracker alignment is best achieved by applying a correction in the track reconstruction. This procedure will be described in this appendix, as well as data-driven methods to validate the alignment.

A.1.1 Tracker Alignment at CMS

The subdivision of the CMS tracker into mechanical structures is described in Table A.1 and Fig. A.2. The tracker can be aligned at different levels of precision, for instance:

1. large mechanical structures,
2. layers and discs,
3. ladders and blades,
4. or sensors.

The alignment parameter of these objects are called *alignables*. Typically, the alignment of the large mechanical structures (sensors) corresponds to corrections of $\mathcal{O}(1\,\text{mm})$ ($\mathcal{O}(10\,\mu\text{m})$). The degree of precision, i.e. how many alignables are considered, is related to size of the sample of tracks that will be used to perform the alignment.

Table A.1 Structure of the CMS tracker and characteristics of the mechanical structures. (Complementary diagram in Fig. A.2)

Acronym	Full name	Substructures	# layers/discs
PXB	PiXel Barrel	2 half cylinders	3
PXF	PiXel Forward	2 × 2 half disks	2
TIB	Tracker Inner Barrel	2 half barrels	4
TOB	Tracker Outer Barrel	2 half barrels	6
TID	Tracker Inner Disk	2 full discs	3
TEC	Tracker End-Caps	2 full discs	9

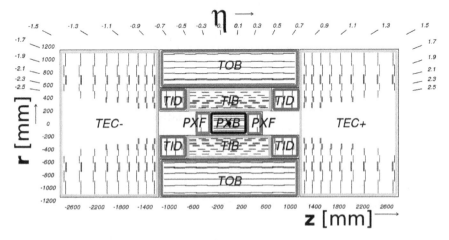

Fig. A.2 Structure of the tracker as of 2016. The pixel (strip) tracker is the innermost (outermost) part. The pixel tracker is made of two layers in the forward region (PXF) and of **three layers in the barrel region** (PXB); the strip tracker is made of four layers in the inner barrel (TIB), of three layers in the inner disk (TID) on each side, of six layers in the outer barrel (TOB) and of nine layers in the end-caps (TEC) on each side. Figure modified from [1]

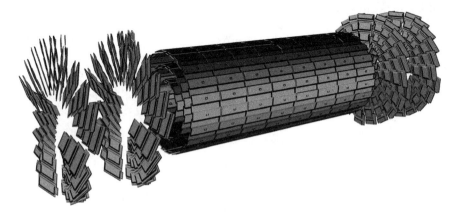

Fig. A.3 The ladders (barrel region) and the blades (end-caps) in the pixel tracker [2]

Fig. A.4 MILLEPEDE-II logo

A.1.2   Track-Based Approach

The main approach chosen at CMS is the track-based approach.[1] It consists in a least-square minimisation of the following fit [4, 5]:

$$\chi^2(p, q) = \sum_j^{\text{tracks}} \sum_i^{\text{hits}} \left(\frac{m_{ij} - f_{ij}(p, q_j)}{\sigma_{ij}} \right)^2 \quad\quad (A.3)$$

where

– p stands for the alignment parameters and q for the track parameters,
– m stands for the measurements and f for the predictions,
– and σ stands for the uncertainties from the measurement.

The difficulty consists in minimising this χ^2 with potentially millions of parameters.

At CMS, two algorithms are used: MILLEPEDE-II [6] and HIPPY (previously HIP) [7]. Both are based on a linearisation of the χ^2:

$$\chi^2(p_0 + \Delta p, q_0 + \Delta q) = \sum_j^{\text{tracks}} \sum_i^{\text{hits}} \left(\frac{m_{ij} - f_{ij}(p_0, q_{0j}) - \Delta p f'_{ij}(p_0, q_{0j}) \Delta q_j f'_{ij}(p_0, q_{0j})}{\sigma_{ij}} \right)^2$$
$$(A.4)$$

The linearisation of the χ^2 allows to treat the problem with linear algebra, as a matrix-inversion problem:

$$C \times (\Delta p\ \Delta q) = b \quad\quad (A.5)$$

Unlike the unfolding problem, standard problem in physics analyses, the problem of this matrix inversion is rather the size of the matrix than instabilities, with a number of rows and columns of the order of several millions.

---

[1]The laser-based method was also performed during Run-I [3]. However, it can only align large mechanical structures.

A.1.2.1   MillePede-II

MILLEPEDE-II is a project developed in Hamburg (originally at the university, and currently at DESY) [6]. External to CMS, it is also used in other experiments, e.g. Belle [8].

The approach consists in performing a global fit, at the same time determining the correction to the alignment of the modules and releasing the track parameters. Therefore, all correlations may be treated in a mathematically rigorous way.

We first re-write the χ^2 in a general form:

$$\chi^2(A) = \sum_k \left(\frac{(m_k - d_k^T \cdot A)}{\sigma_k} \right)^2 \tag{A.6}$$

At the minimum:

$$\underbrace{\left(\sum_k \frac{(d_k \cdot d_k^T)}{\sigma_k^2} \right)}_{\equiv C} \times A = \underbrace{\sum_k \frac{m_k}{\sigma_k^2} d_k}_{\equiv B} \tag{A.7}$$

The large number of parameters can be treated thanks to the special structure of the matrix C. Indeed, it can be partitioned into blocks for *local* and *global* parameters, related to tracks and modules and denoted with Greek or Latin letters, respectively:

$$\text{residuals} = \underbrace{\sum_{i=0}^n a_j \cdot d_j}_{\text{global parameters}} + \underbrace{\sum_{i=0}^\nu \alpha_j \cdot \delta_j}_{\text{local parameters}} \tag{A.8}$$

A priori, given N measurements, the computing time would be proportional to $(n + N\nu)^3$, which grows very fast and would not be achievable with a large number of tracks. But with a bit of block matrix algebra, one can further reduce the computing time:

$$\begin{pmatrix} \sum G_i & \cdots & \Gamma_i & \cdots \\ \vdots & \ddots & 0 & 0 \\ \Gamma_i^T & 0 & L_i & 0 \\ \vdots & 0 & 0 & \ddots \end{pmatrix} \begin{pmatrix} a \\ \vdots \\ \alpha_i \\ \vdots \end{pmatrix} = \begin{pmatrix} \sum b_i \\ \vdots \\ \beta_i \\ \vdots \end{pmatrix} \tag{A.9}$$

(where the letters L and G are explicitly chosen for *local* and *global*, while the Γ, which corresponds to the letter G in Ancient Greek but looks like an upside-down L, relates the two worlds). The trick is to remember that at the end of the day, we are only interested in the global parameters. Therefore,

– if we define the *local solutions* $\alpha*_i$:

$$\alpha_i^* \equiv \mathbf{L}_i^{-1}\mathbf{b}_i \tag{A.10}$$

which corresponds to the solution where the Γ_i matrices are neglected, involving only a small matrix inversion;
– if we redefine the source term:

$$\mathbf{b}' \equiv \sum \left(\mathbf{b}_i - \mathbf{L}_i\alpha_i^*\right) \tag{A.11}$$

which does not require any time-consuming operation;
– and if we define the inverse of the Schur complement:

$$\mathbf{C}' \equiv \sum \left(\mathbf{G}_i - \Gamma_i\mathbf{L}_i^{-1}\Gamma_i^\mathsf{T}\right) \tag{A.12}$$

which involves the same small matrix inversion as above,

then we can redefine the problem with a smaller matrix to invert:

$$\mathbf{C}'\mathbf{a} = \mathbf{b}' \tag{A.13}$$

where the matrix $\mathbf{C}'$ has n rows and n columns. This also means that the time-consuming step is not significantly affected by the use of larger samples: one can increase the statistics to several tens of millions of measurements.

When aligning the pixel detector only, or when aligning the mechanical structures only, i.e. when the $n \lesssim 2000$, the matrix can still be exactly inverted in a reasonable time (few hours). Whenever possible, one *should* use the inversion, since it provides the *exact* solution within the linear approximation.

However, aligning the whole tracker at sensor level, which is a typical example, would require too much computing power: a second simplification is necessary. This simplification is using the MINRES-QLP algorithm [9], which addresses the matrix inversion as an iterative process, minimising $||\mathbf{C}'\mathbf{a} - \mathbf{b}||$ at each iterations. The number of iterations is usually not larger than 2000.

In this context, a full alignment of the CMS tracker with the data of a whole year requires a running time of the order of a day.

In practice, "MILLEPEDE= Mille + Pede":

Mille   determination of all the values needed to calculate the global χ^2, implemented within CMSSW;

Pede   determination of local (track) refits to construct the linear equation system and determination of global (alignment) parameters, implemented in a standalone executable in Fortran 90.

We will see later in this document how to steer and run MILLEPEDEat CMS.

A.1.2.2   HIPPY

The HIPPY project was the first adopted technique of alignment of the CMS detector and is based on a slightly different idea [7]. The fundamental idea is still to start from a linearised χ^2, and to use linear algebra, but the dependence on the track parameters is removed (local-fit approach); to compensate from this assumption, tracks are refit afterwards and the whole procedure is then iterated. Since it is not demanding in terms of memory, it can be used with very large statistics.

At DESY, the activity related to alignment are only focused on MILLEPEDE-II; therefore, HIPPY is mainly mentioned for completeness, but will not be discussed further.

A.1.3   Samples

A natural choice consists in taking Minimum Bias (MB) tracks, in order to scan most regions of the phase space. Additional types of tracks may however be considered.

First, *isolated muons* are used to fill regions of higher transverse momentum in the phase space, where the Minimum Bias (MB) statistics is too low; in other words, since the transverse momentum is higher, the tracks are straighter, which allows to constrain modules in different series.

In the next subsection, we shall see why and how two other types of tracks are crucial to perform the alignment:

1. tracks of *cosmic rays*,
2. tracks of muons coming from the decay of a Z^0 boson (i.e. $Z^0 \longrightarrow \mu\mu$).

A.1.4   Weak Modes

A Weak Mode (WM) refers to any transformation such that $\Delta\chi^2 \approx 0$. Equivalently, it is a transformation that changes a set of *valid* tracks into another set of *valid* tracks. This may happen in our case for two reasons:

1. all collision tracks come from the center of the detector, and the collision are performed in the centre-of-mass frame with identical protons;
2. the detector is symmetric around the beam axis, with respect to the transverse plane passing through the center of the detector, and with respect to the center of the detector.

A.1.4.1   Types

Different types of WMs exist, as, for instance:

(A) Longitudinal view of the tracker illustrating a   (B) Transverse view of the tracker illustrating a twist
telescope deformation. A consequently distorted cos- along the beam axis with the two true (distorted) out-
mic ray is represented.   going muons of the decay of a Z^0 boson represented
                                                      with a plain (dashed) line.

Fig. A.5 Illustration of WMs and their impact on certain track topologies when only MB tracks
are used. **Layers silicon tracker** (tracks) are represented in black (red)

Telescope   longitudinal deformations:

$$\Delta y = C \times z \qquad\qquad (A.14)$$

Twist   azimuthal deformations:

$$\Delta \phi = C \times z \qquad\qquad (A.15)$$

Sagitta   radial deformations:

$$\Delta y = C \times r \qquad\qquad (A.16)$$

The telescope and twist WMs are illustrated in Fig. A.5.

A.1.4.2   Fix with Tracks of Various Topologies

The solution consists in using tracks of cosmic rays and tracks of muons decaying
from a Z^0 boson.

Cosmic rays. These tracks have a different topology than the collision tracks. Therefore, they break the centered symmetry of the problem.
$Z^0 \longrightarrow \mu\mu$ **data**. The two muons coming from a Z^0 boson introduce a momentum
scale in the alignment. Moreover, they connect modules in different directions.[2]

---

[2]Similarly, one can also include $Y \longrightarrow \mu\mu$ data: not only they also introduce a scale, but they also
connect modules in additional directions, as the muons decaying from a Z^0 are mostly back to back,
while muons decaying from a Y meson are rather making a right angle; they may therefore solve
additional WMs. In 2016, they were not ready in time to be used by MILLEPEDEbut were used with
HIPPY.

A.1.5 Time Variations

The alignment is a very fine calibration that can also vary over time. This may be a difficulty, as datasets (especially for cosmic rays) may have very limited statistics.

Time variations come from various origins.

A.1.5.1 Types and Reasons for Variations

Magnet cycles. Ideally, the magnet should be turned on once for all. However, for maintenance reasons, even in data-taking periods, it may be temporarily turned off.[3] This mostly affects the large mechanical structures, for $\mathcal{O}(1\text{ mm})$.

Temperature variations. In data-taking mode, the tracker is cooled down to $-15\,°C$. Also for maintenance purposes, like a long shut-down of the detector, the cooling operations may affect the alignment, this time at the level of the modules, for $\mathcal{O}(10\,\mu m)$.

Ageing of the modules. As we discussed in Sect. 3.2.2.1, the modules operate in a high-radiation environment; therefore, their performance vary with time. In particular, the Lorentz drift, i.e. the drift of the released particles in the modules, is significantly affected and requires dedicated calibration. The Lorentz angle depends on several parameters:

1. the magnetic field,
2. the electric field,
3. the mobility of the charge carriers,
4. the thickness of the active zone.

While the magnetic field is kept constant over time, the three latter parameters are changing with time, due to the high radiation, i.e. the measurement of the position of the hit changes with $dx \approx \tan\theta_{\text{Lorentz}}$, as is illustrated on Fig. A.6. At CMS, modules are mounted pointing either inward or outward (see Fig. A.3), therefore this change over time applies with different signs in the measurement of the position of the hits. The effect is continuous over time, and for a long period of data taking, it has to be taken into account. There is also a dedicated calibration of the pixel modules to compensate for the variation of the Lorentz drift, but it was found in practice that it is optimally corrected when treated complementarily in the alignment procedure, as will be shown later on.

---

[3]Especially in 2015, the magnet was suffering of a severe issue, and was turned on and off quite frequently, forcing the alignment crew to perform the alignment on an almost daily basis.

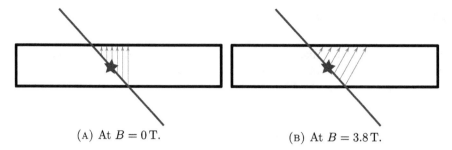

(A) At $B = 0$ T.   (B) At $B = 3.8$ T.

Fig. A.6 Sketches illustrating the *Lorentz drift* due to the ambient magnetic field in which the modules operate. The **module** is in black; the track is in blue; the hit is shown by a red star; and the Lorentz drift is represented by the grey arrows

A.1.5.2   Strategy

Align separately:

– *absolute* positions of the High-Level Structures (HLSs) *with time-dependence*;
– *relative* positions of the modules to the HLSs *without time-dependence*.

The HLS can be

– either the large mechanical structures (already described in Fig. A.2),
– or disks and layer (also represented in Fig. A.2),
– or the ladders and blades (sketched in Fig. A.3).

This strategy turns out to be a good compromise to include time dependence while keeping large statistics, especially for cosmic rays and dimuon data. The choice of the HLSs is motivated by the treatment of the Lorentz drift.

A.2   2016 Legacy Alignment

We present the performance of the alignment in 2016. First we detail the configuration; then we show the performance of the detector for a given interval of time.

A.2.1   Configuration

The period of data taking is divided into 36 intervals of time. It is performed at the level of the sensors, which is possible thanks to high statistics (see Table A.2).

Several iterations are performed with MILLEPEDE-II and HiPPy in order to cope with the linearisation of the problem in case of large deviations. The choice of the HLSs was set to ladders and blades in order to correct for the non-constant Lorentz angle (since a ladder contains modules pointing in the same direction, the correction to its position can absorb the Lorentz drift).

Table A.2 Statistics in use for MILLEPEDE-II. Note: this configuration and this statistics require 150 GB of RAM and around 30 h per iteration

Dataset	#tracks (M)	Weight
MB tracks	13	0.2–0.3
Isolated muons	53	0.25
$Z \longrightarrow \mu\mu$	32	1.0
Cosmic rays	3	2.5

A.2.2 Performance

The difference between two geometries can be checked on a *Geometry Comparison Plot*. In Fig. A.7, we compare the geometry of the detector during the data-taking and the one after the alignment procedure. Each point represents a module; the colour is related to the high-level structure consistently with Fig. A.2. One can see the movement $Y(\Delta r, \Delta z, r\Delta\phi)$ of a module initially at position $X(r, z, \phi)$.

In the current case, clear differences between the tracker in data-taking and after realignment are seen, but are not sufficient to tell whether the alignment has improved. Data-driven quantities are however checked in order to compare the quality of different alignments. Here, we present the following ones:

- distributions of the median of the residuals,
- performance of the reconstruction of the PV as a function of the track kinematics,
- reconstruction of the mass of the Z^0 boson as a function of the kinematics of the outgoing muons.

As the alignment is performed in 36 interval of times, the performance of the alignment can be studied in each of them. Apart from the survey of the Lorentz drift, only one interval of time with large statistic is sufficient to attest the global performance of the alignment.

For more readability, the geometries in data taking and after realignment are respectively in red and blue in the different validation plots. In addition, for reference, a geometry without misalignment is displayed in green.

A.2.2.1 Distribution of the Medians of the Residuals

The *distributions of the medians of the residuals* are a measure of the local precision. Let us consider a set of tracks:

- Each track is reconstructed for different geometries.
- The hit prediction x'_{pred} for each module is obtained from all other track hits.
- Then the residuals $x'_{\text{pred}} - x'_{\text{hit}}$ is histogrammed for each module.
- Finally, for each large mechanical structure, the median of these residuals is plotted.

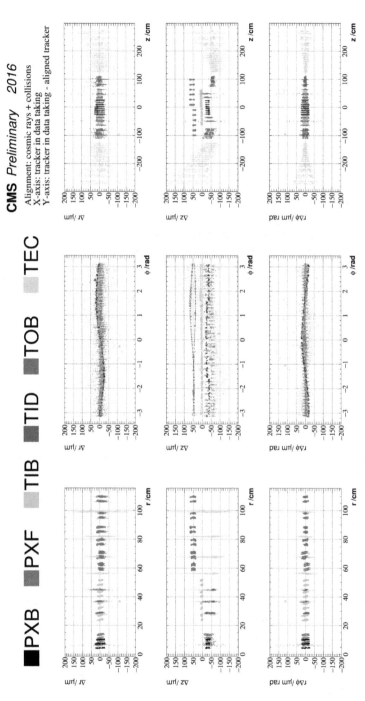

Fig. A.7 *Geometry comparison plot* between the geometry during data-taking and the geometry after 2016 Legacy Alignment

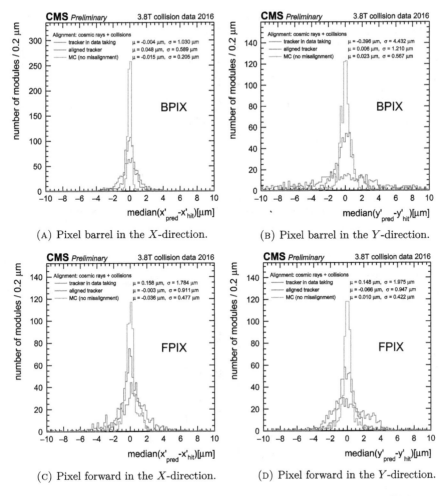

Fig. A.8 Distributions of the medians of the residuals in the pixel part of the tracker [10]

This validation can be seen in our case on Figs. A.8, A.9. Note that in order to avoid statistical correlations, we use independent samples for alignment and validation.

An optimally aligned detector should be peaked around zero, as the Monte Carlo reference suggests. The width has also a component related to the statistics, which is why the Monte Carlo reference, though ideal, still presents a width. Deviations from 0 for the mean indicate systematic biases.

Lorentz drift. The variation of the Lorentz drift can be investigated over time from the distributions of the median of the residuals. In particular, they are produced for each interval of time independently for inward and outward pointing modules. For a wrong correction of the Lorentz angle (typically in the case of a constant geometry), two peaks ($\Delta\mu \neq 0$) will appear in the distributions, as the local x shows in different

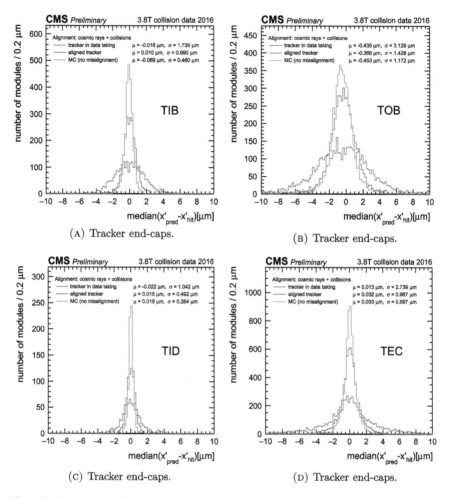

Fig. A.9 Distributions of the medians of the residuals in the strip part of the tracker for 2016 Legacy Alignment [10]

directions for different module orientations; on the other hand, with a new alignment, this would be absorbed by the geometry, as ladders are considered as high-level structures, in which case no double-peak structure should be seen ($\Delta\mu = 0$).[4] This is illustrated on Fig. A.10.

---

[4]It is also possible to disentangle the Lorentz angle from the alignment by using data with magnetic field on and off, but this was not done in 2016.

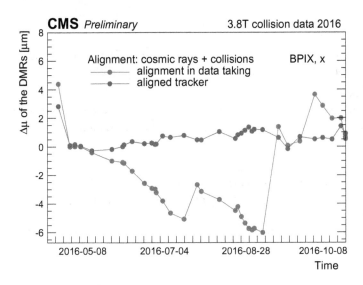

Fig. A.10 Correction of Lorentz drift for the 2016 Legacy Alignment [10]. Difference of means in the distributions of the medians of the residuals over time for inward- and outward-pointing modules in barrel pixel part of the detector. The jumps in the red curve correspond to updates of the calibration of the pixel

Fig. A.11 Picture of the PV validation. Given a vertex composed of N tracks, it is refitted with $N - 1$ tracks; the impact parameter of the Nth track is then studied [11]

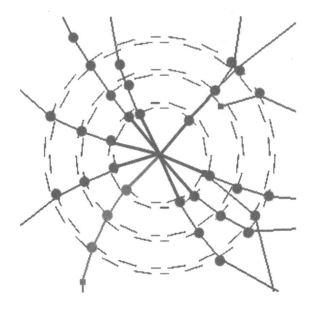

A.2.2.2   Primary-Vertex Validation

The PV tests the alignment by looking at the reconstruction of the vertices.
Let us consider a vertex reconstructed from N tracks. Then

1. consider one tracks;
2. refit the vertex with the $N - 1$ other tracks with the usual reconstruction;
3. and check the impact parameter of the track under scrutiny with respect to the vertex as a function of the direction of the tracks.

Random misalignments translate into an increase of the spread, and simply lead to lower precision, while systematic misalignment translate to biases in the mean (the exact pattern depends on the misalignment), and may lead to systematic biases in the measurement.

To perform the PV validation, we consider tracks from MB events satisfying the following requirements:

Vertex —   at least four degrees of freedom in the vertex fit,
Tracks —   at least six hits in the tracker, of which at least two in the pixel detector, at least one hit in the first layer of the Barrel Pixel or the first disk of the PXF
$\chi^2_{\text{track}}/\text{n.d.f.} < 5$

The case of the 2016 Legacy Alignment is shown in Fig. A.12. In particular, one can see that the modulation in ϕ in the geometry during data-taking is cured after realignment.

A.2.2.3 $Z \to \mu\mu$ Validation

Distortions in the geometry may degrade the kinematics of the two outgoing muons coming from the decay of a Z^0 boson. The reconstruction of the Z^0 boson is thus investigated by measuring its mass as a function of the kinematics of the muons.

The Z^0-boson mass is reconstructed with a Voigtian function[5] with fixed decay width for the Breit–Wigner component, while the background is reconstructed with a exponential function. The mass is then estimated from the mean of the Voigtian function as a function of different variables:

– the azimuthal angles $\phi_{\mu\pm}$ of each of the muons,
– the rapidity separation $\eta_{\mu+} - \eta_{\mu-}$,
– the cosine of the angle of the boson $\cos\theta_{\text{CS}}$ in the Collins-Soper frame.

---

[5]Convolution of Gaussian and Lorentzian functions:

$$V(x) = \int_{-\infty}^{+\infty} \exp\left(-a\left((x - t) - x_0\right)^2\right) \times \frac{1}{b + (t - x_0)^2} \, dt \qquad (A.17)$$

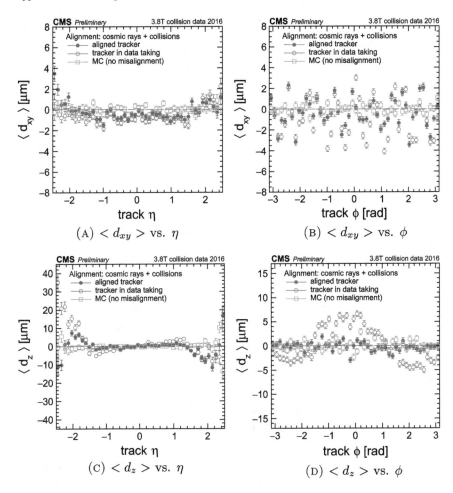

Fig. A.12 PV validation for the 2016 Legacy Alignment, performed with PV tracks [10]. The impact parameter of the track under scrutiny of a given vertex is plotted along the beam axis $< d_z >$ and in the transverse plane $< d_{xy} >$ as a function of its azimuthal angle ϕ and pseudorapidity η. Here, the ϕ modulation and high-η deformations in $< d_z >$ have clearly been fixed

This is shown in Fig. A.13. In addition, a fit of the mass is performed for each geometry.

The selected muons must satisfy the following requirements:

- $p_T > 20\,\mathrm{GeV}$
- $|\eta| < 2.4$
- $80\,\mathrm{GeV} < M_{\mu\mu} < 120\,\mathrm{GeV}$

Note that muons are reconstructed with both the tracker and the muon system, but only the geometry of the tracker is updated in the validation.

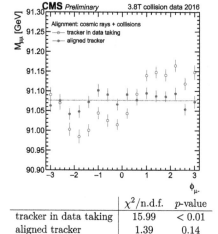

	χ^2/n.d.f.	p-value
tracker in data taking	15.99	< 0.01
aligned tracker	1.39	0.14

(A) Z^0 mass as a function of the negatively charged muon.

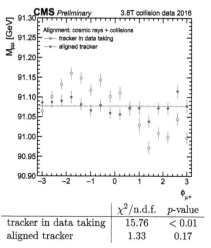

	χ^2/n.d.f.	p-value
tracker in data taking	15.76	< 0.01
aligned tracker	1.33	0.17

(B) Z^0 mass as a function of the positively charged muon.

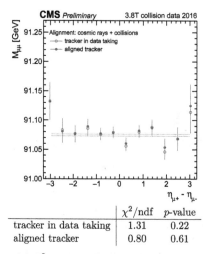

	χ^2/ndf	p-value
tracker in data taking	1.31	0.22
aligned tracker	0.80	0.61

(C) Z^0 mass as a function of the pseudorapidity separation in the Collins-Soper frame.

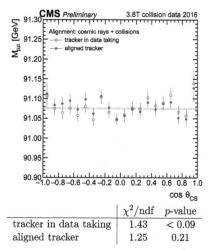

	χ^2/ndf	p-value
tracker in data taking	1.43	< 0.09
aligned tracker	1.25	0.21

(D) Z^0 mass as a function of the cosine of the angular separation in the Collins-Soper frame.

Fig. A.13 $Z^0 \to \mu\mu$ validation for the 2016 Legacy Alignment [10]. The mass of the Z^0 boson is fitted as a function of different kinematic parameters of the outgoing muons. Any deviations indicates systematic misalignment of the tracker

A.3   Summary and Conclusions

The alignment of the tracker is a crucial step in the procedure of calibration of the detector; in the hierarchy, it is one of the first to perform, and many other calibrations rely on it: calorimeter alignment, muon alignment, calorimeter calibration, beamspot calibration, etc. Given the large size of the silicon tracker, dedicated techniques needed to be developed; these were concisely presented. As an illustration, the alignment campaign of the 2016 data was presented in this appendix; it is one of the most precise campaigns ever performed at CMS so far, not only in terms of data, but also in terms of configuration.

Other aspects of the alignment, e.g. online automated calibration or other campaigns, are also covered in Ref. [12].

References

1. Chatrchyan S et al (2014) Description and performance of track and primary-vertex reconstruction with the CMS tracker. JINST 9(10):10009. https://doi.org/10.1088/1748-0221/9/10/P10009, arXiv:1405.6569 [physics.ins-det]
2. Chatrchyan et al S (2010) Commissioning and performance of the CMS pixel tracker with cosmic ray muons. JINST 5:T03007. https://doi.org/10.1088/1748-0221/5/03/T03007, arXiv:0911.5434 [physics.ins-det]
3. Sirunyan AM et al (2017) Mechanical stability of the CMS strip tracker measured with a laser alignment system. JINST 12(04):04023. https://doi.org/10.1088/1748-0221/12/04/P04023, arXiv:1701.02022 [physics.ins-det]
4. CMS Collaboration (2010) Alignment of the CMS silicon tracker during commissioning with cosmic rays. J Instrum 5(03):T03009. http://stacks.iop.org/1748-0221/5/i=03/a=T03009
5. The CMS collaboration (2014) Alignment of the CMS tracker with LHC and cosmic ray data. J Instrum 9(06):06009. http://stacks.iop.org/1748-0221/9/i=06/a=P06009
6. Blobel V, Kleinwort C (2002) A new method for the high-precision alignment of track detectors. In: Proceedings of the conference on advanced statistical techniques in particle physics (2002)
7. CMS Collaboration (2018) The HIP algorithm for track based alignment and its application to the CMS pixel detector. http://lib.tkk.fi/Diss/2007/isbn9789521037115/article6.pdf. Accessed 07 Feb 2018
8. Abe T et al (2010) Belle II technical design report. arXiv: 1011.0352 [physics.ins-det]
9. Choi S-CT, Paige CC, Saunders MA (2011) MINRES-QLP: a Krylov subspace method for indefinite or singular symmetric systems. SIAM J Sci Comput 33(4):1810–1836
10. CMS Collaboration (2016) CMS tracker alignment performance results. https://twiki.cern.ch/twiki/bin/view/CMSPublic/TkAlignmentPerformanceEOY16. Accessed 08 Feb 2018
11. Musich M Primary-vertex validation. Private communication
12. Mittag G, CMS Collaboration (2017) Alignment of the CMS tracker: latest results from LHC run-II. J Phys: Conf Ser 898(4):042014. http://stacks.iop.org/1742-6596/898/i=4/a=042014

Index

© Springer Nature Switzerland AG 2019
P. L. S. Connor, *Inclusive b Jet Production in Proton-Proton Collisions*, Springer Theses,
https://doi.org/10.1007/978-3-030-34383-5

CPSIA information can be obtained
at www.ICGtesting.com
Printed in the USA
LVHW050618151220
674155LV00006B/15

9 783030 343859